Additives

Your Complete Survival Guide

EDITED BY
Felicity Lawrence

CENTURY
LONDON MELBOURNE AUCKLAND JOHANNESBURG

Notes on Contributors

Felicity Lawrence joined Haymarket Publishing in 1982, and has been editor of *New Health* magazine since 1984. She studied classics at Oxford.

Geoffrey Cannon is co-author of the best-selling books *The Food Scandal* and *Dieting Makes You Fat*. His third book *Fat to Fit* is published this year. He wrote the *Sunday Times* features which first revealed that the NACNE report on dietary goals for Great Britain had been suppressed. He studied philosophy, psychology and physiology at Oxford.

Caroline Walker is a nutritionist and started the City and Hackney Health Authority heart disease prevention programme. She studied biology at London University and nutrition at the School of Hygiene and Tropical Medicine. She is a member of council of both the Coronary Prevention Group and the London Food Commission. She is co-author of the bestseller *The Food Scandal*.

Melanie Miller studied biochemistry and environmental science and obtained an MSc at Aston University. She has spent three and a half years researching additives for her thesis, and in 1985 she produced the report *Danger: Additives At Work*. She now works at the London Food Commission.

Dr Peter Mansfield is a family doctor in Lincolnshire. He studied medicine at Cambridge and London. He founded the Templegarth Trust, with the aim of finding practical ways to promote positive health in modern communities. He is author of several pamphlets and a book about health, *Instinct For Life*, and drew up a pamphlet for the Soil Association, *Look Again At The Label*, a consumers' guide to additives.

First published in 1986 by Century Hutchinson Ltd, Brookmount House, 62–65 Chandos Place, Covent Garden, London WC2N 4NW

Century Hutchinson Publishing Group (Australia) Pty Ltd, 16–22 Church Street, Hawthorn, Melbourne, Victoria 3122

Century Hutchinson Group (NZ) Ltd, 32–34 View Road, PO Box 40-086, Glenfield, Auckland 10

Century Hutchinson Group (SA) Pty Ltd, PO Box 337, Bergvlei 2012, South Africa

Set in Linotron Ehrhardt by
Rowland Phototypesetting Ltd, Bury St Edmunds, Suffolk
Printed and bound in Great Britain by
Richard Clay (The Chaucer Press) Ltd, Bungay, Suffolk

ISBN 0 7126 1269 6

Contents

Acknowledgements

This book has grown out of a series of articles which originally appeared in *New Health* magazine, and thanks are due to Haymarket Publishing and everyone at *New Health* for their support. The Food Additives Campaign Team has also inspired us and the authors would like to thank all its members. Adriana Luba devoted long hours to trailing round supermarkets to draw up the product lists on pages 237 and 243, and especial thanks are due to her.

While responsibility for the judgments in this book is ours, our thanks go to: Sue Dibb, Bill Horner, Dr Sandra Hunt, Dr Tim Lang, Dr Alan Long, Dr Erik Millstone, Naheed Mehta, Julie Sheppard, Professor John Soothill and Dr Fred Steward.

Our publishers worked hard to produce the book to a tight schedule, and thanks go to Claire Brown, James Crocker, Gwyn Lewis, Gail Rebuck, Sarah Riddell and Sarah Wallace.

Professional support has been given by: Matthew Bullard, Bill Hamilton, Alexandra Henderson, Carole Hobson, Jacqueline Korn, Deirdre McQuillan and Deborah Rogers.

Felicity Lawrence
1986

1 How to Use This Book

FELICITY LAWRENCE

'I would like to take this opportunity to reassure anyone worried by the recent publicity, that all permitted additives are safe', Mrs Peggy Fenner, Minister of State for Agriculture, Fisheries and Food (MAFF) at the opening of the Fast Food Fair in Brighton in November 1985.

'If that is the MAFF view, we have a government department guilty of criminal complacency', Barry Sheerman MP, at the launch of the Food Additives Campaign Team in December 1985, responding to a representative from the Ministry of Agriculture who insisted that all additives were thoroughly tested and known to be safe.

Who should you believe? Where does that leave you, the consumer? Can you trust government to protect your best interests before those of the food industry? Does any of the fuss concern you at all?

We are all exposed to additives. Have a look at some of the labels on your food. If you eat a fairly typical breakfast of fruit juice, cereal, toast and jam and coffee, you could easily be consuming thirty to forty doses of additives. Add a cup of tea and a couple of biscuits mid-morning, and you could be getting another dozen or so. Then perhaps you have a sandwich lunch, clocking up another thirty doses of additives. An afternoon snack, a quick and easy to cook supper of frozen meat and veg, and consuming a cocktail of 150 additives in a day would be nothing unusual. On page 237, there is a list of common foods which would provide you with just that. And those are only the additives which have to be declared on labels; anything which is used as a 'processing aid' need not appear, so, for example, bleach used to whiten floor need not appear on the label on your loaf of bread.

Maybe you are already careful about what you eat. Even so additives are almost impossible to avoid – they even make their way into and on to many foods which have a healthy image – dried fruit, wholemeal bread and yoghurt to name but a few. Unless you are obsessional about avoiding additives, you too will be getting your daily cocktail of chemicals added to food.

There has been an explosion in both the number and volume of additives used in the last few decades. It is estimated that over 200,000 tonnes of additives are used a year in this country – that's approximately eight to ten pounds of additives per person. The quantity has increased ten fold in the last thirty years and industrial sources expect the market to continue growing rapidly. Over three-quarters of the food we eat has been processed in some way. No one knows exactly how many additives there are, but reliable estimates put the figure at approximately 6,000, with the great majority of those being flavourings.

Mrs Fenner may have given us her reassurances in 1985, but the fact is that less than ten per cent of the additives in use in this country are controlled by permitted lists. Members of the food industry sit on the government bodies which make recommendations on additives, and have enormous influence over the way these chemicals are approved and regulated. Consumers have had little say. Does any of this matter, or should we just accept it as part of the great march of technological progress?

As the use of additives has grown, so too has the evidence of ill health either directly or indirectly attributable to the chemicals in our food. Eczema, asthma, rashes, headaches, gastric irritation, fits, diarrhoea, hyperactivity – reports of adverse effects suffered as a result of eating additives are becoming more and more common as doctors begin to recognise the problem (see chapter 6). The long-term effects of regular consumption of additives are not known; the risks of cancer, birth defects and genetic damage are poorly understood (see chapter 7).

In assessing additives, government weighs the risks against the benefits. But the risks are borne by the consumer, while the benefits nearly always fall to the manufacturer. And even establishing what the risks are is often a major problem. Current methods of safety testing have disturbing shortcomings, and those who are responsible for conducting them are among the first to acknowledge their limitations (see chapter 3). Additives are effectively presumed innocent until proven guilty, and the proof has to come from consumers. A number of the additives permitted in this country are banned abroad because of the hazards they present, and there have been many instances in the past of additives which were thought to be safe by government being banned when they were later discovered to be unsafe.

Government often ignores warnings of danger from its expert committees. Industry lobbies have bent the ear of the regulators for decades, while consumers have been virtually unrepresented. You will find that our children and workers in food industry are in effect human guinea pigs in a huge experiment with our food whose outcome may not be known until it is too late (see chapters 5 and 8). Britain has been called the sickest country in Europe. Our rates of premature deaths from heart disease and cancer are among the highest in the world. Our rates of premature deaths from all other causes are the highest in Europe. The quality of our diet has deteriorated and additives are used to disguise that poor quality.

With the help of this book, you individually can do a lot to protect both yourself and your family and to bring about change.

We each have to face a chemical burden today which is unprecedented. We are exposed to harmful chemicals from agricultural, industrial and urban pollution – add to that the burden from additives in food, and it may be the last straw. Our beleaguered immune systems cannot cope. But by reducing the toxic load wherever we can, we give our bodies a chance to cope. The occasional dose of an additive in food is not going to kill you. You will not get cancer just because you have consumed one suspect additive. But the more you are exposed to harmful chemicals, the greater your risk of suffering ill health.

By using the charts at the back of this book (pp. 110–236), you can reduce your personal toxic load considerably. Once you have learnt how to read and understand food labels (p. 107), you can use the numerical guide to E numbers and other additives to avoid the most suspect ones. The ratings given to each additive enable you to tell at a glance which are hazardous. You can use the Guide to Safer Shopping (p. 243) to find products without suspect additives quickly. You will discover that many additives are only added because of other additives in a product. Once you have understood the function of additives, you will know how to look for fast and convenient food which is nonetheless free from harmful chemicals.

Every time you make a more informed choice, you will reduce the toxic burden your body has to cope with. You can make sure that your children are not exposed to the worst hazards and sensitised when they are young. You can also learn how to judge which foods offer real nourishment and the vitamins, minerals, essential fats and fibre we need for health. And by choosing more nourishing foods you can strengthen your ability to cope with any harmful additives you cannot avoid. Studies with animals show that those which are better fed are less susceptible to the harmful effects of additives; a diet rich in vitamins will strengthen your resistance to chemical onslaught, and consumption of fibre can reduce the potential damage of additives.

You can also bring your own personal pressure to bear on the situation. As a result of consumer pressure some sectors of the food industry are already changing their policies on food additives. By writing to retailers and manufacturers, by lobbying your MP, you can influence the pace of change. And whenever you refuse to buy poor quality food containing suspect additives you help to bring us all closer to securing a healthier food supply.

2 Legalised Consumer Fraud

CAROLINE WALKER

Most discussion about additives focuses on their toxicity, but there is another side: consumer fraud, and it too has a profound effect on our health.

When you read the list of ingredients on a packet of food, forget food, think chemistry. Ask yourself, why is this flavouring or colouring in this particular product? Why does it need emulsifiers, thickeners, anti-caking agents and a plethora of other peculiar additives? As soon as you start to think about additives in this way, with the help of the charts (pp. 110–236), you will begin to understand just what a load of rubbish you are being offered to eat by some food companies. Others, of course, try to do a good job. In the face of increasingly difficult odds, as more and more manufacturers pollute their products with cheap ingredients and artificial flavours and colours, some manufacturers are sticking to real food. We should give them our support. But we should also be aware that an increasing number of foods on sale in our shops are tarted up by additives, fancy packaging and advertising, pretending to be something they are not. And nearly all of us eat them, one way or another.

Most public discussion about additives has focused so far on their poisonous attributes – whether they cause a range of physical reactions from spots to cancer. It is right for us to be concerned about the direct effects of additives on health (and they are covered in detail in later chapters), but there is another side to the discussion which is usually neglected. It is the issue of consumer fraud and the legalised debasement of food, both of which in turn have a profound effect on our health.

Additives and Your Health

We in Britain are at the top of the first division in the league tables for heart attacks in people under sixty-five years of age. We also have the dubious honour of leading Western Europe in rates of early death and illness (before sixty-five years of age) from all other causes. We are in terrible trouble. Every year preventable illnesses cost the country billions of pounds in lost working days. More billions are spent on the health services as doctors try to patch up the results of a lifetime's (or half a lifetime's) lousy nutrition, smoking and lack of exercise. Constipation, piles, diverticular disease, middle-aged diabetes, overweight – by the age of forty, quite apart from tooth decay, at least half of us suffer from one or other of these diseases, caused in part by the food we typically eat in this country.

What is wrong with our meals and what has this to do with food additives? Processed starches, sugars and fats; that's what's wrong. Years of eating white bread, cakes, biscuits, puddings, sweets, sauces, sausages, pies, soft drinks and other wonders of food technology have ruined the nation's health. Instead of living on healthy fresh fruit and vegetables, wholemeal bread and cereals, fresh fish and lean meat, we live off the supermarket shelf and eat out of packets, a habit we learned in the nineteenth century with the introduction of food science's earliest inventions – sweets, cakes and biscuits. It is true that we no longer die in thousands from chronic starvation. But we suffer and die in more subtle ways, and our daily meals are considered by medical authorities the world over to be the underlying cause.

Additives are used to embellish artificially hardened (saturated) vegetable and animal fats, processed sugars and starches, to turn them into a multitude of 'different' foods. In practice, what you get nine times out of ten is a product lacking, or very low in, vitamins, minerals, essential polyunsaturated oils and fibre, all crucial to good health. And it's all within the law, for large manufacturers have perverted the course of justice within Whitehall and Westminster, and persuaded civil servants to draw up legislation for their benefit, and MPs to nod it through.

Today's food legislation makes it perfectly legal to sell substandard food, to pass off rubbish as a delicacy. And it is additives which make it all possible.

Modern Miracles

New twentieth century foods are the inventions of some of the best brains in the UK. An imaginative food scientist, with a flair for sculpture and design and an understanding of the latest inventions in machinery, can create an endless array of textures, shapes, flavours and colours, and present them to us as cakes, biscuits, noodles, snacks, sweets, drinks, puddings, sauces, soups. All of them are made with varying proportions of processed starches, sugars and fats and the odd fragment of real food, and they are stuck together with – you've guessed it – additives.

Processing aids smooth the passage of ingredients through machinery, having 'purified' raw materials of tiresome relics of their past life which wouldn't fit into the recipe. Such things as plant fibre, essential polyunsaturated oils which go rancid, natural colour and flavour – all of these can be removed with solvents, machines and processing aids to leave behind the stuff the food scientist is really after – purified starches, sugars and fats, all with reliable viscosity, texture and tastelessness.

These basic ingredients can then be mixed with emulsifiers (to keep the fats well dispersed), stabilisers, firming agents, gelling agents, aerating agents, anti-caking agents, texture improvers, thickeners, thinners, binders, buffers. Water and air will stand up to order. Crisps will crinkle, drinks fizz, sauces shine, nothing is impossible.

Then comes the really artful stage: the creation of what is known in the trade as 'organoleptic' qualities, meaning 'sensory' to you and me. This is where some 6,000 flavours (nobody knows the exact number) come into their own. For flavour manufacturers boast that they can create absolutely any flavour demanded, from hedgehog crisp to Bombay duck. A dash of colour to give visual appeal and this marvel of technology is now ready to visit the packaging department for the final touches.

First, a brand name suggestive of country life, like Granny Goodbody's Country Fresh Traditional Olde English Pie Mix. But it must also, according to the law, have a description in keeping with its ingredients. So in the small print you find: 'Apple and Blackberry Flavour Pie Mix – just add milk and eggs' (often the only ingredients of any nutritional value must be added by the purchaser). Second, a wrapper designed to appeal to every proud housewife in the land. Throw in a nationwide advertising campaign and it's certain to be a real winner.

Within the food industry brilliant minds are constantly at work, dreaming up ways to spread out the expensive bits of the recipe, add cheap substitutes, replace real food with artificial colour and flavour – and so degrade the product. No need to put real egg in custard when you can get away with the artificial colours E_{102} (tartrazine), E_{110} (sunset yellow) and E_{127} (erythrosine) and emulsifiers and starch to thicken it instead, not to mention a tasty artificial flavouring or two.

Lemonade contains no lemon, cheese and tomato snack no cheese. In 1978 23,000 tonnes of cheese went into manufacturing. In 1983 the quantity was almost halved to 13,700 tonnes. Why? Cheese analogues, that's why. Where there's a will, there's a way. Where there's cheese, there's a cheese analogue (the clever way of saying artificial cheese).

A packet of chicken soup has the following list of ingredients, which, by law, are given in descending order of weight: modified starch, dried glucose syrup (glucose? syrup? I thought this was the first course, not the pudding), vegetable fat, dextrose (more sugar), flavour enhancers: monosodium glutamate, sodium 5'-ribonucleotide, salt, dried chicken (real food at last!), flavouring, onion powder, caseinate, acidity regulator (E340), spices (luxury! real flavour!), emulsifiers (E471, E472(b)), colours (E150, E102), antioxidants (E320, E321). Phew!

Has it ever occurred to you that the flavour enhancers are enhancing the flavour not of real ingredients like chicken, but of artificial flavours? A cheese and tomato flavour snack contains cheese and tomato flavour, flavour enhancer (621, otherwise known as monosodium glutamate which is prohibited in food for babies and young children), and flavouring. Note how the manufacturer lists 'cheese flavour' separately from the other 'flavourings', giving the impression that the snack contains real cheese, which it doesn't. If it did, you can bet the manufacturer would say BIG VALUE – WITH REAL CHEESE! all over the front of the packet to make sure that you appreciated what you were buying.

If you think this is an exaggeration of what goes on behind the doors of food factories in the 1980s, here is the list of ingredients in Raspberry Flavour Trifle:

Raspberry flavour jelly crystals: sugar, gelling agents (E140, E407, E340, potassium chloride), adipic acid, acidity regulator (E366), flavourings, stabiliser (E466), artificial sweetener (sodium saccharin), colour (E123)

Raspberry flavour custard powder: starch, salt, flavourings, colours (E124, E122)

Sponge: with preservative (E202), colours (E102, E110)

Decorations: with colours (E119, E132, E123, E127)

Trifle topping mix: hydrogenated vegetable oil, whey powder, sugar, emulsifiers (E477, E322), modified starch, lactose, caseinate, stabiliser (E466), flavourings, colours (E102, E110, E160(a)), antioxidant (E320)

Disgusting, I hear you say. I never eat those things. Well, someone does.

Foods, if that is the right word, like this are a nutritional disaster. Stripped of essential nutrients, they supply virtually nothing but calories. They represent the legalised debasement of food. The chemicals which make it all possible are passed as safe and necessary by Parliament, thanks to the constant and skilful lobbying which goes on in Whitehall to make sure that new legislation helps the manufacturer.

The Ministry of Agriculture sidesteps the issue of nutritional quality (or lack of it) in products like Raspberry Flavour Trifle by pointing out that food is part and parcel of the enjoyment of life; nutritional value is just one aspect of the function of our daily bread. So, they say, let manufacturers innovate and improve. Provided it is safe (whatever that means) and they label it, it's a good food. Consumers can choose.

Illegal Fraud

How has our food legislation got into such a mess? It is an interesting question, because when our first food laws were drawn up in the nineteenth century the government of the day could have been in no doubt that consumers needed protection. For in the eighteenth and nineteenth centuries wicked wholesalers and retailers got away with murder, literally, poisoning their customers with metallic colours used to embellish poor quality food. Bread was adulterated with alum, chalk, pea flour, potato flour, pipe-clay and even powdered flints. 'Secondhand' tea was processed in eight factories in London, where used tea leaves (purchased from hotels and coffee-houses) together with hedgerow clippings, were carefully dried and coloured with lead, copper, indigo, Prussian blue – anything to make them look like tea.

The manufacture of beer, a basic part of the nineteenth century diet, using cheap substitutes for malt and hops, became an established art, to the extent that brewers' guides were published and republished, giving details of the delicacies an enterprising brewer might use to cheapen the recipe. Beans, berries, coriander, caraway, alum, gypsum and, of course, extra water (the cheapest substitute of all) were all part of 'normal' beer manufacture.

Chicory and acorns went into coffee, copper colouring into pickles, bitter almonds into wine, red lead into the rind of Gloucester cheese, floor sweepings into pepper, metallic colours into sweets and cakes. The meals of the mass of the British population were so heavily adulterated it was reliably reckoned by the analysts who monitored this appalling situation to be impossible to buy pure, clean food. Why did it happen?

People who live in the countryside, and live off the land, know where their food comes from. In a small community where everyone knows the miller, the baker and the brewer, and furthermore where local assizes or laws control the quality of basic commodities, there are incentives for food producers to maintain standards. There can be little room for fraudulent practice, because sooner or later the naughty person will be found out. So until the mass migration of the population into the towns and cities of nineteenth century Britain, people knew what they were eating. But when they moved into the cities all that changed.

Gone were the local miller, baker and brewer; gone were the days of local barter and the pig in the back garden and the home-grown cabbages and root vegetables. An urban population found itself at the mercy of unscrupulous manufacturers and retailers who lost no time in profiteering from their weakness and ignorance about mass food production and distribution. Removed from the site of food production and living in poverty, they had no bargaining power. And worse, there were no laws to prevent consumer fraud.

In 1820 a book emblazoned with a skull and crossbones on the cover was published. *Treatise on Adulterations of Food and Culinary Poisons* was a gripping account of widespread fraud in the manufacture of British food. Written by Frederick Accum, a respected analytical chemist, it was an instant bestseller. With its quotation on the front 'There is Death in the Pot', it attracted attention from every paper and periodical, and must have been the subject of fertile debate at dinner parties as the guests picked their way through chalk-infested bread and doctored wines.

For the next fifty years a long and hard battle was fought by Accum and fellow-chemist John Mitchell, by Thomas Wakley, editor of the medical journal *The Lancet*, and the physician Arthur Hassall among others, to persuade government to legislate against adulteration. These campaigners had an enormous public following.

They finally got what they wanted in the shape of the Sale of Food and Drugs Act 1875, which made it an offence to sell food which was not 'of the nature, substance or quality of the article demanded'. In other words, a loaf of bread had to be what it said it was; no more, no less. But this successful legislation was not won without tremendous effort, which should be a lesson to all of us today. For the manufacturers who put these 'additives' into food said that they improved food, that the public wanted food made like that, and that if it was all so terrible people wouldn't buy it. Furthermore, they maintained that their 'additives' actually enhanced taste and appearance, and that it was impossible to produce pure food at a price people were prepared to pay. Years were wasted as the government dithered over the choice between statutory or voluntary controls. Worse, lives were wasted as the public ate their way through poisonous adulterants and were served with food of substandard quality.

The lesson from the nineteenth century could not be clearer. Adulteration of food was practised because it was more profitable. People bought bad food because they had no choice. There was no effective legislation to stop it. The health of the people suffered.

Legal Fraud

How well has the quality of our food been protected since the dark ages of Victorian Britain? The food legislation we have today is based on the Sale of Food and Drugs Act 1875. Under the 1984 Food Act it is an offence to: 'add any substance to food, use any substance as an ingredient in the preparation of food, abstract any constituent from food, or subject food to any other process or treatment, so as to render the food injurious to health.' And if a person sells any food not of the nature, substance or quality demanded, they are guilty of an offence. Moreover, manufacturers and retailers must obey the Trades Descriptions Act, food labelling regulations and the volumes of food law relating to individual products. So far, so good.

Standards Flag

The Sale of Food and Drugs Act 1875 was in time followed by Food Standards for different products, further safeguards to ensure good quality, many of which exist to this day. Thus, margarine cannot contain more than 16 per cent water, fish cakes must contain a minimum of 35 per cent fish (a pretty miserable amount when you think about it), white flour by law consists of 72 per cent of the whole grain (the inner 72 per cent), salad cream contains a minimum oil content of 25 per cent. And so on.

About ten years ago an interesting but little publicised change of attitude took place. The Ministry of Agriculture made a fundamental shift in their attitude towards Food Standards, which culminated in the abolition of the Food Standards Committee (FSC) in 1983.

For years the FSC had closely monitored the quality of British food products and had advised government about new Food Standards and other legislation. More recently, under the chairmanship of Professor A G Ward, FSC had produced some excellent reports on subjects such as meat products, margarines, table spreads, cheeses, water in foods, misleading claims and labelling. Their recommendations, if implemented in their entirety, would have done a great deal to help consumers choose good quality food.

In their 1979 report on food labelling, FSC examined the problems for consumers of the proliferation of convenience foods in our shops, caused by the use of both new processing techniques and new ingredients. It was agreed that Food Standards could not possibly be drawn up for all foods; there were too many of them. 'The amount of legislation necessary would clearly outweigh any resultant benefits.' However, FSC said that Food Standards should be kept for basic staple British foods, like bread and meat products. But how could the customer be protected against the proliferation of near alternatives, which were not covered by compositional standards? For example, as soon as a piece of legislation dictated that a Fish Finger must contain 70 per cent fish, an enterprising manufacturer intent on selling a less fishy product could simply alter the shape a bit and change the name to Fish Fiddle. FSC's answer to this problem was: improve food labelling.

They said: 'We would not wish to criticise the search for substitute and modified foods which are nutritionally adequate and safe to eat. Provided these foods are labelled, advertised and promoted for what they are, as required by law, there can be no objection to their sale. Indeed, it is our firm belief that controls imposed on food manufacturers should wherever possible be such that innovation and technological developments can still be used to full advantage for the benefit of both consumer and manufacturer. However, some of these developments could result in substandard versions of commonly known and well recognised foods.' So proper labelling is needed.

There was another reason why Food Standards were not seen to be so important. The Ministry of Agriculture, Fisheries and Food (MAFF), along with the medical profession and the Department of Health (DHSS), resolutely refused to acknowledge (until the 1984 COMA (DHSS) report on diet and cardiovascular disease was published) that a substantial amount of illness in Britain was due to British food. No one was keeling over from starvation; in fact many of us were too fat. So obviously further Food Standards for nutritional reasons were unnecessary.

Lastly, Food Standards impede trade. Britain had entered the EEC and was committed to the general principle of harmonisation, to facilitate free trade with her neighbours. Food Standards were out. For a few traditional foods like butter, cheese, mustard, corned beef and orange squash they would stay. But for the rest, manufacturers could innovate and improve. Provided they label food and we like it and it is 'safe', they can sell it and we can eat it.

So that seems quite simple. Label food and we will all know what it contains. Or will we? FSC has made a number of recommendations to government over the years, many of which have been ignored. For example, it said that the percentage of added water in all foods should be declared, so should mechanically recovered meat and reformed meat (x per cent); artificial flavour should be declared 'artificial', the quantity of lean meat in meat products should be given (x per cent), ingredient lists should be headed 'ingredients in weight order', all cheeses should be labelled with their fat content . . . The list goes on; MAFF continues to ignore them.

Labelling is only useful if we understand what is written, if it does not muddle and deceive us. With today's sophisticated processing techniques and new ingredients whose names we do not understand, is labelling really enough? Do you think manufacturers should be allowed to bung into their cooking pots constipating starches, artery-clogging fats and tooth-rotting sugars (often labelled with peculiar names), carefully pepped up with additives to add a little zing to the mixture? The problem is that without compositional standards for the majority of foods, manufacturers are free to use almost any ingredients they like. With legislation that permits artificial flavourings and colours to be tipped into most mixtures in almost limitless quantities, with permission to use processing aids which between them allow the creation of seemingly endless variety, with food labelling laws that provide for no immediately obvious distinction between real and artificial, and with sophisticated packaging and advertising to make you believe almost anything, today's consumer is in a mess.

Most discussion about the use of food additives completely misses the point about consumer deception. Indeed, most 'experts' on additives appear never to have thought about the issue. Their minds are focused firmly on two issues, and the giveaway clues are: potatoes and preservatives.

At conferences on food safety and quality, you know what you're in for when the lecture on additives starts like this. Food scientist strides confidently on to platform. Demands first slide. Up comes picture of potato, whereupon lecturer delivers brisk commentary on poisonous chemical – solanine – *naturally occurring* in humble spud (potatoes are a real favourite for natural poisons). Nervous titters from the audience. Second slide: picture of a miserable slave in Ancient Egypt dipping meat into vat of preserving salts . . .

Potatoes and preservatives tend to come up wherever discussions about food additives take place, particularly in food industry magazines, university food science departments and in government reports. The argument goes: additives are necessary, otherwise we would all be keeling over with food poisoning. Foods 'naturally' contain poisons, therefore food has never been totally safe. What's more, preservatives have been used ever since man stopped being Neanderthal, so what's all this argument about? Eat up your dinner and stop fussing.

This is a preposterous attitude. First, if food does contain natural poisons, that is no reason for adding more. Rather it is an argument for extreme caution in permitting the use of any additives, so as not to increase the toxic load.

Second, arguments in support of preservatives are of no relevance to any other additive. And of the thousands, literally, of additives used in the UK, just a handful are preservatives. They account for less than one per cent (by weight, number and value) of all additives used. All the rest are flavours, colours and processing aids, which are of greater benefit by far to the manufacturers than to the consumers. How?

Added Value

How can it be in a manufacturer's interest to spend so much time inventing such revolting recipes as the Raspberry Flavour Trifle (p. 14)? One answer is that the manufacturer who can provide the cheapest product obviously gets the retailer's contract which, with the expansion of multiple retailers, becomes more and more crucial to manufacturers' survival. Now if one factory dreams up a recipe for raspberry trifle without any raspberry in it, using cheaper ingredients instead, provided the retailer can flog it to the public (with a bit of help from advertising, of course), then the raspberryless trifle rapidly becomes the rule rather than the exception.

The other answer is Added Value. All good businesses have to expand, and that means new investment in machinery and technology, to keep up with competitors. It also means new products.

Put yourself in the shoes of Sam Sludge, managing director of Sludge International plc, creator and sole purveyor of the Bulldog Brand Boil-in-the-Bag Soyburger, Iron Lady Wonder Whip and Honey Crunch Rainbow Jelly Toppings. Last year's profits up, surplus cash to spend. Sludge Puddings reports good trade in the previous year. So do Sludge Snacklets and Sludge Sweeties. So rather than write out a hefty cheque to the Inland Revenue you pay a visit to your top secret Sludge Laboratories where Professor Crackling is putting the finishing touches to his latest creation, a new instant chocolate flavour pudding created with the lovely cheap ingredients found for him by Sludge International Research and Development Division, which constantly scours the globe in search of cheaper and better raw materials to turn into new and profitable foods.

'A complex multiphasic hydrocolloidal system of water, lipids, protein, carbohydrate and air', mutters Professor Crackling, tingling with excitement and doing a little skip as he deftly tips the brown powder into cold milk, does some energetic whisking with the Sludge Whisk-o-Pud (yours for just twenty Sludge Pudding Packet tops) and artfully swirls the resulting light and creamy fluff into his best cut-glass test tube for an experimental mid-morning snack.

'Perfect!' he proclaims, savouring every last molecule of TFS/P/84/28, his new chocolate pudding flavour, 'Organoleptic bliss! Now I can create twenty other puddings exactly the same, but they will all be different!' For there is no tiresome chocolate in the mixture to prevent the basic recipe being turned into peach, pear, strawberry or his newest tropical tutti-frutti fizzy flavour: all the flavour is artificial. A quick shake of reliable artificial yumminess and the mixture can be endlessly transformed. All it needs is a good supply of processed sugars, starch, saturated fat, a few by-products of the dairy trade (nice and cheap courtesy of the EEC's ludicrous farming subsidies) and the all-important additives to make the whole thing work. Emulsifiers to keep the fat and water properly mixed and make air stand up (that's where the whisking comes in), gelling agents to encourage solidification, antioxidants to encourage immortality and stop the fats turning rancid when the packet sits on the shop shelf for weeks on end, flavour enhancers to enhance the flavour of the artificial flavours, and a dash of colour for visual appeal. Sam Sludge could not ask for more. All it needs is some clever packaging and advertising, and it'll be ready to send out on a promotional tour.

Added value has nothing to do with added *nutritional* value, far from it. Subtracted nutritional value would be a better description. No, added value is all about *capital* value. It's about increasing the profits from basic raw materials by turning them into more and more different foods. The more foods a manufacturer can create from the same basic ingredients, the better for business.

In the 1950s the British population was expanding. There were more mouths to feed, which meant the food industry could grow in size. Today, the population is static, or possibly declining in size, which is a disaster for food companies. There is a limit to the quantity of food we can be persuaded to eat. And a further problem is that, because of our chronic lack of activity, we are actually eating less per head, than in the 1950s. Hard to believe, but true. We eat fewer calories today than we used to, so profits clearly have to be made in another way.

Look at potatoes from the point of view of the nation as a whole. Potatoes sold as such don't do much for the wheels of commerce. Turned into crisps, on the other hand, they work wonders for suppliers of machinery, oils, additives (colours, flavours, antiox-idants), packaging, transport and advertising. 1,000 lbs of potatoes make just 270 lbs of crisps, because 100 lbs go in the bin from peeling, 27 lbs are lost in trimming, 18 lbs in slicing and 667 lbs (mostly water) in cooking. But add in 82 lbs of oil and flavouring, and there is a product with added capital value par excellence. If we ate all of the potatoes produced in the UK as potatoes, none of this industrial activity would exist. And if we ate all the potatoes as potatoes, it would cost us a lot less, and give us a lot more nourishment. Next time you visit the shops, check the price of a bag of crisps and compare it with an equal weight of potatoes. You'll find yourself wondering if you got the arithmetic right.

Below are just a few examples of legalised consumer fraud.

The Wet Meat Movement

Why does bacon spit and splutter the minute it hits the frying pan? Added water. With the aid of polyphosphates, clever manufacturers have worked out how to sell you the cheapest item in the food technologists' handbook: water. Water is added, quite unnecessarily but within the law, to bacon, ham, sausages, hamburgers, chickens, fish, fish fingers and a host of other meat and fish products. Water is cheaper than meat or fish.

When the 1984 Meat Products legislation was being drawn up, civil servants were subjected to extremely strong pressure from manufacturers, who were anxious to ensure that they could get the maximum quantity of water into bacon and ham before it had to be declared as 'added water' on the packet. They succeeded. The amount of water that can now be added legally to cured meats before a declaration must be made is 50 per cent more than is needed to effect the cure (which is done by injecting water and curing salts into the joint). That is why bacon and ham are dripping wet: legalised consumer fraud.

And it's why fish and fish products are also getting wetter than wet. Trading Standards Officers are increasingly concerned at the quantity of water added to fish and fish products, much of it with the aid of additives, and the FAC is even now considering the problem. But will it listen as attentively to the consumer's voice as we all would like?

Cheap fats too are good fillers. Together with water, they work wonders for a meat product when better quality ingredients are a bit pricey. But it can't be done without additives to stick it all together and colour it pink. Take a look at the ingredients list of a few meat and fish products, and you'll see what it's all about.

Flavour, Flavoured

A prawn flavour crisp contains no prawns. That is legal. The law states that a manufacturer can use the word 'flavour' when all the flavouring is artificial. There is no requirement to say 'artificial flavour' or 'imitation flavour' or 'synthetic flavour', all of which we would immediately understand. No, all a manufacturer has to do is find a way to dispense with real, fresh ingredients and put in some tasty imitation flavour instead. Think of the advantages: no mess with perishable raw food, no storage or transport problems, just some clean, reliable artificial flavouring. Add in a bit of extra filler to bulk up the product (milk by-products like whey are a favourite, or cellulose, sugars, fats) and there is an item that's good for business. The Food Standards Committee of MAFF actually recommended in 1979 that the words 'contains artificial flavour' should be written on the packet. Their recommendation was never implemented.

The word 'flavoured', on the other hand, indicates that some real ingredient has been used. Peach Flavour Yoghurt contains no peach. Peach Flavoured Yoghurt contains some peach, but exactly how much they do not tell you because there is no requirement to give the quantities of ingredients used. Peach Yoghurt, on the other hand, contains even more peach, but again you have no idea how much.

Legalised consumer fraud again. Why should manufacturers be allowed to hoodwink the public in this way? It is just as bad as the practices of the nineteenth century, only this time it is within the law.

Vitamin Vitality

Another way of tarting up poor quality ingredients to make you think you are getting something good is to add minerals and vitamins. Think of the number of breakfast cereals that advertise themselves as containing 'added vitamins and iron'. Most of them are made of processed cereals from which the valuable fibre and germ have been removed, so it's not surprising that vitamins have to be added to restore their respectability. However, do not assume that the 'vitamin-added' variety is as good as the real thing. They seldom add back all the minerals and vitamins they took out.

'Health' drinks, too, which consist chiefly of water, sugar, flavour and colour, are doctored with added vitamins, usually vitamin C. Any drink that consists mostly of water and sugar cannot be good for health, with or without added vitamin vitality. So why are they allowed to be advertised as 'health' drinks?

Consumer Power

Is there anything you can do about the quality of British food? Most of us tend to assume that the individual is powerless against industrial giants. And up to a point that may be true. But food companies can only survive if we buy their products. All of us have to eat and, en masse, the population could have enormous clout to persuade government to tackle the issue of consumer fraud. So write to your MP, talk to the shop manager, write to retailers' and manufacturers' headquarters (the address is on food packets), and, most effectively of all, refuse to buy rubbish. After all, it's your health and your family's health that is at stake.

3   Tests . . . What Tests?

MELANIE MILLER

*Additives then are used to tart up processed fats, sugars and starches –
foods which contain few of the vitamins, minerals, essential fats and fibre
we need – and as such they contribute to a major health problem. But
surely additives in themselves can't be dangerous. Governments wouldn't
allow them if they weren't safe . . . or would they? The fact is that current
testing of additives is at best inadequate, at worst irrelevant. Where doubts
have been raised about the safety of additives, government response is
usually to call for more tests and allow us to continue eating them while
they wait a few years for the results.*

We tend to assume that experts know best, that government wouldn't
allow all these additives in our food if there was any doubt about their
safety. Unfortunately, such faith is misplaced.

Innocent Till Proven Guilty

Take just two examples. In 1980 the government body responsible
for assessing the safety of additives looked at modified starches. It
listed thirteen modified starches for which it said further evidence
was needed to quell doubts about their safety. It also recommended
that a 'permitted list' of modified starches be drawn up. 'We would
emphasise that unless the further evidence requested is produced
within three years from the publication of this report, we shall feel
bound to recommend that the substances in question be removed
from the permitted list.' It also recommended that 'starches modified
by the use of epichlorohydrin and/or propylene oxide should not be
permitted in foods described either directly or by implication as being
specially prepared for infants and young children.' Both these agents
are suspected of causing mutations and cancer. That was six years
ago. What has happened about modified starches? Have the suspect
ones been removed? No, 'The Food Advisory Committee is still
considering representations being made by interested parties,' a
Ministry of Agriculture representative said recently. For 'interested
parties' read industry.

In 1979 a government report on colourings said more research was needed to establish the safety of seventeen colourings. If satisfactory results were not published within five years, they should not continue to be allowed in food. That was seven years ago. And where are those colourings? Shelved until the evidence is good enough to show that they present no danger to consumers? No, the only place you will find them shelved is in your local shops, for they are still permitted for use in food. Rather than temporarily banning them or severely restricting their use, the regulators continue to permit these additives in food while they wait for the results of safety tests. Additives appear to enjoy the privilege of being innocent until proven guilty.

Many additives never stand trial at all. Flavourings, which make up the great number of additives used by industry, have never been subjected to systematic testing, and there are no regulations restricting their use. Part of the justification given by government for not testing all flavours is that they are used in such small quantities that they cannot pose much risk to consumers. Most flavours *are* used in relatively small quantities; but some, such as ethyl vanillin, are very widely used, and some people may be eating them from a variety of sources in significant doses. Further, the fact that flavours are so potent at low doses indicates they are likely to be relatively reactive chemicals and therefore may well pose a health hazard.

In fact, some of these unregulated additives are already known to be harmful substances. Dr S Gangolli, a toxicologist at the British Industrial Biological Research Association (BIBRA), the government and industry-funded research organisation, has said: 'There are some [flavours] that could be potentially a problem', because they have structures which are similar to chemicals known to be harmful. Some flavours are already known to pose hazards (see chapter 4).

The Notion of Need
Why is industry still allowed to use these additives which have not been tested or are acknowledged to be suspect? How do they get approval?

Responsibility for assessing additives is split between two government committees, answerable to the Ministry of Agriculture, Fisheries and Food and the Department of Health. The Food Advisory Committee (FAC, formerly more explicitly called the Food Additives and Contaminants Committee, FACC for short) assesses additives on the basis of two criteria: need and safety. Acceptable proof of need covers everything from whether the additive in question enables industry to reduce its costs or to introduce technological innovation, to whether there is some supposed consumer demand for the additive. The FAC is required to restrict the use of additives to the minimum, but interprets this very loosely. The UK permits more additives than other European countries; for example, in this country we apparently 'need' seventeen azo dyes, while the rest of the EEC copes with a range of twelve, and Norway's consumers and industry manage perfectly well without any.

When we joined the EEC, instead of restricting our lists, the government fought for special permission to continue using the additives that had not been approved and were not needed by the other European countries.

Who Does the Tests?

Once the FAC has been satisfied that some sort of need can be demonstrated, it submits the additive to the Committee on Toxicity, known as COT, which then looks at what evidence there is on the additive's safety.

Tests on new additives are carried out by the companies which want to produce them, either in their own research laboratories or in those of a contract laboratory or BIBRA, which they pay to do the testing for them. Testing is very expensive – anything from £10,000 to £500,000 or more may be needed, depending on how many problems are noticed with the additive. How wide a variety of tests is done depends on how much the manufacturers think the additive will be used, whether humans have been exposed to it in some form before and how toxic they think its chemical structure might be. The test results are assessed first by the manufacturers; they then submit their report to COT for consideration. Their reports are submitted in confidence, and much of the evidence reviewed by COT is not published and so not available for public scrutiny.

The World Health Organisation/Food and Agriculture Organisation (WHO/FAO) and EEC also sponsor committees made up of industry representatives, scientists and government officers to review additives, and COT looks at their reports too. Sometimes COT will refer an additive to another specialist UK committee on a point of particular internal controversy – for example, if it is not clear whether an additive is a mutagen (i.e. capable of causing damage to genes), the Committee on Mutagenicity may be asked to comment.

Having looked at the evidence, such as it is, COT classifies additives in one of several groups: group A additives are those which it thinks are safe, group B are those which it only accepts provisionally, and on which it has asked for more safety data within a specified time if they are to continue to be used, groups C and D are those which ought not to be allowed in food, group E are those for which the available evidence is inadequate and group F those for which there is no information.

Then it's back to FAC, which in turn reviews the additive and makes a recommendation to ministers that the additive either be allowed, restricted or banned.

Recommendations Ignored

But ministers can choose to ignore the evidence of the expert committees if they want. Once a recommendation has been made, interested parties are consulted and make representations. Ministers may be subject to intense lobbying by the manufacturers, who after all may have already invested considerable resources in developing a new additive, or who have a lot at stake if an established and widely used additive is about to be banned. In practice, recommendations of the expert committees have often been ignored (see chapter 4).

So much for the structure of government decision-making on additives. Food industry is responsible for testing, and quite naturally the primary object of its research is to produce enough safety data to get an additive passed by government. Unlike most other EEC countries and the USA, the UK government does not commission independent research. It says the FAC invites evidence from all interested parties, including the general public. But this offer is worthless if consumers do not have the resources for independent testing and evaluation. Moreover, even when its advisory committees call for more safety testing or express doubts about any particular additive, the government may simply ignore its advice.

Even if government did accept its committees' representations, even if other groups carried out the research rather than the manufacturers with a vested interest, how valuable would the safety tests be? Are they relevant to real life, are they even scientific? Many toxicologists, the people who actually do the testing, now admit that the safety testing of additives is inadequate and inappropriate.

The Cocktail Effect

A major fault with the testing procedure is that additives are tested singly, but they rarely find their way into consumers' stomachs in such splendid isolation. In real life they are eaten in complex combinations, with all sorts of possible interactions. Last Christmas, for example, the Food Additives Campaign Team presented Mrs Thatcher with a hamper of what the average family might eat for its Christmas dinner: it contained no fewer than 170 doses of additives.

In a rare test carried out on two additives at once, sodium sulphite (E221) and benzoic acid (E210), both common preservatives and often used in combination, it was found that their effects on health were more pronounced when they were mixed than when they were tested singly. The same was found to apply to mixtures of aspartame and monosodium glutamate. Additives are rarely used singly in foods; an average processed product is likely to contain four to seven additives, many contain far more. A single meal may contain an elaborate cocktail of thirty or more different additives. These cocktails of additives may react with each other and with food to produce new chemical substances of unknown toxicity. It is these new products which should be tested for safety, not just the individual additives. But they are not. From current tests, it is just not possible to judge whether these mixtures are safe yet that is what industry and government do. It would be very expensive and time-consuming to test all possible combinations, as industry quite rightly points out. But that doesn't mean we have to accept things as they are. There is an alternative: to reduce the number and levels of additives used, as an EEC consumer handbook, *Food Additives and the Consumer*, points out: 'The possibility cannot be ruled out of two substances, both harmless by themselves, interacting to yield a product which is toxic. Even apart from any toxic potential, the obvious limits of our knowledge therefore militate in favour of a reduction in the numbers of permitted substances; there are many doctors who would like [permitted additive] lists to be as short as possible.'

Even the editor of the trade journal *Food Manufacture* recently admitted: 'Maybe the number of additives in food should be reduced.' Yet there has been no recent move within the UK to reduce our lists of permitted additives and they remain longer than those of other European countries.

Animals Are Not Humans

Another fundamental criticism of safety tests is that additives are tested on animals, and animals are not the same as humans. Toxicologists themselves have criticised the value of these tests in relation to humans. Professor Conning, director of the food industry-funded British Nutrition Foundation, and former director of BIBRA (which carries out many of the tests), has himself pointed to 'the difficulty of predicting long-term consequences by using tests of relatively short duration. For example, man may consume small amounts of any given component daily for some seventy years, but it is impractical to undertake tests for longer than a few years at most'.

So assumptions have to be made about how exposure to an additive in the short lifetime of the species being tested, usually rats or mice, relates to exposure in the human lifespan. But this isn't really possible. Again Professor Conning points out: 'None of the animal models exhibit the range of disease entities that characterises ageing man.'

What's more, many of the effects of additives are not things you can detect in animals. You can't ask a rat how it feels. An additive may make it feel sick, give it a headache, a stomach ache, make it experience tingling sensations or induce mild behavioural changes or just make it feel generally unwell. But unless the rat actually is sick or starts chewing up its mates or shows some obvious physical or chemical change, you have no way of knowing.

Large mammals, such as dogs or monkeys can give a much better idea in some ways of the effects of a chemical on humans than rats or mice. But to get a proper statistical sample, large numbers of animals have to be tested. Dogs and monkeys cost a lot to keep, and so the cheapest and therefore commonest solution is to use rats and mice. Problems in interpreting tests are made still worse by the fact that the numbers of animals used in practice can still be too low to give statistically useful results.

How do you in any case extrapolate from the effects of an additive on one species such as rats to humans who are a quite different species? Our biochemistry and genetic make-up is different: we may not be affected by additives which are poisonous to other species, but more seriously for us, we may well be affected by additives which other species can tolerate. Thalidomide is a prime example of the difficulty of extrapolating from one species to another: humans turned out to be over a hundred times more sensitive to the drug than rats and twenty times more sensitive than monkeys. In tests on some azo dyes at very low levels of exposure, humans were found to be five times more sensitive to damage than rats.

The problems with animal testing don't end there. Animals used in tests are specially bred, and often inbred. It is easier to duplicate results with inbred animals: for example, testers can work out how high the incidence of tumours would be in the animals normally and so how far tumours in the animals being tested should be attributed to the additive. Humans, however, are not, generally speaking at least, inbred. Men and women exhibit a wide variety of genetic combinations, and therefore a wider range of possible reactions, than a group of homogenous rats.

Now that it is recognised that nutritional status affects results of tests, animals in laboratories are generally fed a good diet with all the vitamins, minerals and other nutrients they need. They also live in a rarified environment. Humans, in sharp contrast, are subject to a whole range of stresses and pollutants. The people who eat most additives are liable to be precisely those who do not get the vitamins, minerals, essential fats and fibre they need for health, and the way additives affect them is likely to be different as a result.

Odds Against

There are endless problems in translating results from animal tests to humans. Dr Erik Millstone, a lecturer at the University of Sussex, has pointed out that there is 'wide-ranging disagreement on how to extrapolate from the results of tests on small groups of animals to large groups of human beings who do not live in laboratory conditions. There are at least twelve different competing statistical techniques for this extrapolation, and the results which they produce may disagree by up to four orders of magnitude.' So it's bad luck if the company doing the tests happens to choose the wrong method!

If an additive only produces adverse effects in a limited though substantial number of people it is unlikely to be detected in testing. Say an additive could produce cancer or birth defects in one in 10,000 people. A test on fifty rodents only has a one in 200 chance of showing it up, if at all. And, of course, the odds are only as good as that if the chemical happens to be one which is carcinogenic for rodents as well as for humans in the first place. Many agents which are carcinogenic to humans are not to animals, and vice versa. Even the relatively large numbers of animals used are not enough to predict acute effects that may occur in a minority of people, such as hypersensitive reactions.

Dr Magnus Pyke, a food scientist, has summarised the problems: 'Two statisticians have calculated that even if a substance were sufficiently toxic to produce 100,000 cases of cancer in a population the size of the USA, there is a one in three chance that the biological tests carried out on 1,000 animals would classify it as safe.'

The standard animal tests can only give a limited amount of information about the effects of any additives on human health. They successfully indicate some acute effects – if a group of mice drop down dead when they eat an additive, you know there's a problem. And such substances are generally not approved as additives. But there are immense problems in trying to detect chronic effects of small doses of additives over a number of years, and acute effects, such as allergies, which only affect a minority. A toxicologist from BIBRA has pointed out that, because of the problems with animal tests, it is 'purely by good luck rather than good management' that there has been no record of a consumer dying suddenly after eating a food additive in this country.

It might be expected that government committees concerned with food safety would monitor the exposure and health of food workers, since they provide an early warning of possible hazards to consumers. Some toxicologists have recently advocated monitoring workers, and the fact that no such monitoring is done by the committees responsible for consumer safety is surprising. Given the difficulties toxicologists have in trying to extrapolate the information from animal tests to human experience, their failure to monitor worker exposure can only be taken as an indicator of the lower priority given by government bodies to consumer and worker safety (see chapter 8).

The fallibility of methods for testing additives is all too obvious, and the fact that many additives have been declared safe in the past only to be found to be hazardous and withdrawn later highlights the size of the problem.

Fiddling the Results

Even when tests are done properly, they have limited value. But alarming evidence of fraudulent practices in laboratories came to light recently, casting still further doubt on them. In some cases, animals which were likely to die over the weekend were thrown away on Friday night because they would be too decomposed by Monday for an autopsy. This meant that crucial results from affected animals were completely lost. In another case, animals which escaped from their cages were simply put back in to any cage, so treated and untreated animals were mixed up. Occasionally laboratories have used fewer animals in experiments than they should, and fabricated data – because it saved money.

A glaring case of fraudulent practice was uncovered in the USA. A major testing company, Industrial Bio-Test, was recently closed down because it was found to have published fabricated and flawed data. An independent pathologist, Adrian Gross, found that some of the animals which died during tests were not examined for cause of death, but were instead replaced with fresh animals – the number of deaths recorded as caused by the additive being tested was misleadingly low as a result. He also discovered that the animal cages were not properly secured, and that wild rodents had entered some of them, completely confusing the results.

As a result of the scandal which followed, the US Food and Drugs Administration issued a Code of Good Laboratory Practice, and set up an inspection force. Recently the Department of Health has set up a small inspectorate to encourage laboratories to comply with the good laboratory practice code in the UK too. However, the financial pressures on laboratories to cut corners are still there, and the inspectorate is small. We have no guarantee of good practice for additives tested in the past. Where tests were improperly conducted, the results are likely to be misleading. Industrial Bio-Test has been called 'the Cadillac of the testing industry' and performed one third of the world's safety testing, on which government regulators based many safety evaluations. For all the substances tested by Industrial Bio-Test, and by any laboratory where good practices are not adhered to, the results are worryingly unreliable.

How Much Is Too Much?

The aim of most safety testing is to establish the level at which an additive can safely be eaten. Government committees do this by taking the maximum level at which an additive causes no obvious toxic effect to animals and dividing it by a safety factor of around 100 to calculate a safe dose or 'acceptable daily intake' for humans. Given that statistical analysis of tests can produce results which differ by four orders of magnitude, the possible margin of error in these so-called 'safe' doses is enormous. Hardly very reassuring. Moreover, 'safe' doses are calculated to prevent obvious ill health, not to promote good health. If they were, much more attention would be paid to the effects of additives on vitamins and minerals. And as if that wasn't bad enough, when the expert committees find that the average consumer may well be exceeding the acceptable daily intakes which they themselves have set, they have been known to simply decide that the safety margins must be wrong.

In 1979 the FACC, as it then was, found that the average-sized adult could very easily consume more than the safe dose of caramel colouring made with ammonia. They also found that a child could consume more than five times the acceptable safe dose of the colouring yellow 2G. The conclusion: the acceptable daily dose must have been too low. The fact that the government committees establish an 'acceptable daily intake' for an additive does not mean that it is safe. In the past 'safe doses' have been set for additives which have later been banned, such as ponceau 3R (a red azo dye).

But quite apart from all this, the idea of a 'safe' dose is completely inappropriate in some cases. The formation of cancer is related to dose in a very complex way; in some cases extremely small levels of exposure have triggered enormous and irreversible ill effects. Similarly hypersensitive reactions can be triggered by tiny amounts of an additive. For people who are hypersensitive to an additive like tartrazine (E102), a safe dose does not exist. Tiny quantities of the additive are sufficient to provoke reactions such as asthma attacks.

Calculations of 'safe' doses do not take sufficient account of people's diets and nutritional status. Assumptions based on averages have to be made about what people eat. In practice, some individuals eat much larger quantities of some foods (the ones they enjoy or can afford) than others. In some cases, people eat quantities well above the calculated safe dose: young children, for example, eat a high proportion of processed foods, snacks and sweets, and are likely to exceed the 'safe' dose for some additives as a result.

Whether a consumer is getting adequate vitamins and other essential nutrients in his or her diet also affects the extent to which he or she can cope with potentially harmful chemicals. Ironically, those people on poor diets of additive-laden processed foods will be among those least able to cope with toxic additions. Some people are much more vulnerable to additives – the elderly, children whose immune systems are not yet fully developed, those who have a particular genetic make-up, those who have been ill, those who suffer from hayfever or asthma, those who are under stress are all more at risk, and in practice that means an awful lot of people.

Safety First?

With so many doubts about its effectiveness, you may wonder what the point of additive testing really is. The lack of relevance to humans, the inadequacies of standard tests – it all suggests that tests are not seriously intended to provide information about the safety of additives for humans.

Over thirty years ago two toxicologists, Barnes and Denz, raised the same doubts. They made fundamental criticisms of safety testing methods, and concluded that animal tests were conducted as measures of 'administrative expediency'. More recently, Professor Conning has criticised safety testing methods for additives, pointing out that 'the development of toxicology over the last three decades has been largely at the behest of regulatory demands . . . it has been easier and cheaper to provide the regulatory system with the data it needs, rather than to argue the scientific point.' He also said that: 'Another confounding factor has been the commercial pressure to achieve clearance of compounds through the regulatory obstacles by the easiest route, and never mind the future consequences.'

This kind of pressure has meant that additives with known hazards and additives about which insufficient is known have managed to get approved for use in food. A BIBRA report mentions that, of additives which have government approval, 'approximately 25 per cent fall in . . . [a class of] compounds deserving the highest priority for investigation because of the known toxicology associated with some of their component structures.'

It is hard to study the testing process and look at the number of additives in use which are known to be harmful to health without coming to a simple conclusion: safety is not the first priority of the system. Daphne Grose of the Consumers' Association, who sits on the FAC, commented in 1982 that a number of additives 'have been found to have question marks against them . . .' and concluded: 'there is almost a consensus now that the number of additives has got to be reduced.' And yet there is no sign of this happening in the UK.

Regulators and industry want to have it both ways. When tests on laboratory animals find in favour of additives, they are presented for approval as positively safe for humans; when tests show adverse effects, they are quick to point out the drawbacks of current methods of testing and to argue in favour of continuing to use an additive while further tests are conducted.

In cases where substances have been banned in the past, it has not generally been until overwhelming evidence has accumulated against the additive. Again and again it seems that many additives are assumed safe until there is incontrovertible evidence that they are harmful. The lack of scientific consensus and lack of information becomes an excuse for regulators to succumb to commercial pressures to do nothing. This applies particularly to additives that have been in use for many years. For the sake of consumers, additives should be presumed harmful (and not used) until all reasonable doubts about safety are removed. There are far too many question marks over large numbers of the additives currently approved.

In some cases where additives have eventually been banned, the food industry has been given plenty of time to adjust – at the expense of consumers' and food workers' health. For example, chloroform, suspected for years to be a serious health hazard, was finally banned as a flavour (hardly an essential function) in July 1979. The FACC declared that there was 'no need for this use to be stopped immediately' and advised that 'time should be allowed for industry to reformulate its products and for the sale of existing stocks'. Sale of existing stocks of food containing chloroform was allowed to continue until April 1981. Production of food containing chloroform was allowed until April 1980. The health of the public seems to be the lowest priority.

What Do We Really Need?

Given the enormous problems with testing additives, a much greater emphasis should be placed on the real need for them. The FAC at present only assesses need in the narrowest of terms. Far stricter judgements should be made, taking into account other technologies which are available. Safety and health ought to be the highest, not the lowest, priority.

Ministers are required by law to restrict the addition of substances of no nutritional value to food. The majority of additives have no nutritional value. The FAC sidesteps this strong legal requirement. It appears to be dazzled by sophisticated arguments of technological need put forward by the food industry. Regulators have to weigh up the needs of industry against the risks to consumers. In practice, the interests of industry take priority – and it is consumers who have to bear the risks.

4 Trade Secrets . . .

GEOFFREY CANNON

. . . or, what they don't want us to know about additives. E is only for a tiny fraction of additives. The great majority of food additives are unregulated, undeclared and are not on the label. The work of the experts who advise government on additives and health is covered by the Official Secrets Act. And when is a secret an Official Secret? When it's a trade secret – and that's official. Policy is made by committees on which the giant food manufacturers are prominent. They give their best advice. But is it in our interest? And is it in the national interest?

The Story of Instant Passion

How many additives are there in the food you eat? It's a secret. Take instant snack meals. The one in front of me says 'BEEF AND TOMATO FLAVOUR' on the front. On the back in small letters there is an ingredients list. Here goes: 'Ingredients: wheatflour; vegetable oil with antioxidants E320, E321; beef and tomato flavour (flavouring; flavour enhancers 621, 627, 631; colours E102, E123, E124, E150, 154; acidity regulators E262, E331; artificial sweetener, saccharin); maltodextrin, processed soya pieces, salt, carrots, tomato, preservative E220; sachet, tomato sauce.'

So how many additives are there in a 'beef and tomato flavour' instant snack meal? There's butylated hydroxyanisole and butylated hydroxytoluene (E320 and E321), two antioxidants whose function is to delay fats going rancid, prohibited by law in food 'prepared for babies and young children'. There's monosodium glutamate (MSG), sodium guanylate and inosine 5'-disodium phosphate (621, 627 and 631), three flavour enhancers designed to enhance the flavour of the beef and tomato flavour, also prohibited in food for babies and young children. There's tartrazine, amaranth, ponceau 4R and brown FK (E102, E123, E124 and 154), four azo or coal tar dyes to stain the noodles yellow, red, red again and brown: of these, amaranth, ponceau 4R and brown FK ('For Kippers') were identified in 1979 by a report of the government advisory Food Additives and Contaminants Committee (FACC) as dubious and put on probation pending tests for cancer and other ill-effects on laboratory animals. Results of these studies have never been disclosed. Manufacturers have voluntarily withdrawn all colours from foods for babies and young children. There's caramel (E150), identified in 1979 as a possible cause of gut problems, pending more animal studies. There's sodium acetate (E262), a preservative, and sodium citrate (E331), a versatile additive, both of which are reckoned to be harmless except as two more sources of sodium in an already overloaded food supply. There's saccharin, a probable cause of cancer in laboratory animals, in the USA allowed in food only if the label carries a health warning, and prohibited here in food for babies and young children. And there's sulphur dioxide (E220), a preservative which, like other sulphur compounds, can irritate the gut and skin.

The phrase 'prepared for babies and young children', by the way, is taken by government to mean products identified by the manufacturer as baby food. If a mother brews an instant snack meal for her small child, that's her look-out.

There you have it, then: this 'beef and tomato flavour' instant snack meal contains fourteen additives, right? Wrong. In common with most highly processed food, the ingredients list includes the word 'flavouring'. Depending on who is doing the counting, between 3,500 and 6,000 flavours are used in British food. In 1986 the value of the food flavour market is around £70 million a year, considerably more than any other category of food additive. (Preservatives, for example, account for around one hundredth of the food additive market, by number, volume, or value.)

How many chemical flavours are there in packages of processed food with the word 'flavouring' or 'flavours' on the ingredients label? I cannot tell you. It's a secret. Flavours are not subject to any regulation or control. Manufacturers are free to put any number of flavours, in any quantity, into their products, without making any declaration except the word 'flavouring' or 'flavours' on the label.

Over nine out of ten of all food additives are flavours; and only the technologists who devise flavour mixes for use in processed foods know what they are.

Does this matter? In 1965 the Ministry of Agriculture, Fisheries and Food (MAFF) published a report of the government advisory Food Standards Committee (FSC) on 'Flavouring Agents'. The committee was told by the trade that most flavours are used in very small quantities and that the volume and number of flavours used in food was not likely greatly to increase. The trade 'strongly reject at this time the drawing up of a permitted list of flavouring agents to be used in food'. Think of the trouble and expense! So that's all right, then? In 1965 the FSC thought not: 'We cannot however accept the trade's view that long usage and no apparent ill-effects are sufficient evidence to demonstrate that a substance is harmless in the absence of satisfactory toxicological data.'

How after all can you be sure that a synthetic chemical introduced into food is safe unless it is checked out? The Committee was 'somewhat disturbed that there was really no toxicity data on most of these flavouring agents'. It seemed that: 'in the amounts customarily used in food the flavours are not acutely toxic but no certain indication is given of any possible chronic effects. As we have stated before in consideration of other food additives, any chronic effects would be insidious and it would be difficult, if not impossible, to attribute them with certainty to the consumption of food containing flavouring agents.'

The Committee therefore recommended that in due course all food flavours be subject to regulation. Twelve years later, in 1977, nothing had happened. Daphne Grose of the Consumers' Association was disturbed. Writing in the journal of the Royal Society of Health in October 1977, she quoted a survey carried out in 1971 which showed that nine out of ten people believe that additives should not be allowed until proved safe. It also turned out that most people believe (wrongly) that all additives can only be used in specified foods in fixed quantities.

Speaking up for the consumer, Ms Grose wrote: 'The gap between actual practice and consumer expectations and belief that all additives are tested, controlled, and limited as to use, should be closing. In due course there should be a positive list for flavours. No additive should be permitted for use unless it is on a positive list.'

In which case, like other classes of additives, flavours could be coded and listed on the ingredients lists.

Are expert government advisers and consumer representative groups fuss-pots? The manufacturers of highly processed food think so. So evidently does MAFF.

Can we trust food manufacturers? To put it another way, can the interests of industry and the interests of consumers conflict? Common sense says yes. Take food flavours, bearing in mind that of the thousands now in use, an unknown (secret) number are neither naturally found in foods, nor 'nature-identical'. Say that there are 1,500 artificial flavours now used in our food, synthesised by food chemists and used in complex cocktails. Is it not conceivable that one of these thousands of chemical concoctions is addictive? Goodness me, what a disgraceful and irresponsible idea, you may think.

But consider the following scenario (totally fictitious, of course). Suppose that sales of Hope & Glory plc's 'Froo-Tee Krunch' were slumping and that a market newcomer, 'Olde Mother Maggie May Country-Fresh Instant Bliss' fruit flavour snacks, made by Sunshine Foods, was booming. Would the boffins in the white coats at H & G Organoleptics Division sit on their hands? Of course not. A task force would be set up with a mission: find the secret of Instant Bliss. Batches of the snacks would be taken apart, analysed chemical by chemical and fed to rats. What did the Sunshine boys know that Professor F K Brown, head of research at Hope & Glory (known to his colleagues as 'Kipper') didn't know?

Suppose then that Kipper cracked it, and an 'eyes only' memorandum went to the H & G board. Suppose the Sunshine file said 'tests reveal significant steep-peak monophagic ingestion preference sequences correlated with esters of ethyl 2, 4-hexadienoate', and that Dr Brown turned up at the board meeting with a test-tube full of 'Ingestion Preference Factor', a passion fruit analogue. Eyes bulging with anticipation, Bill Glop, H & G MD, would say: 'Kipper, what does the Sunshine file mean?' The Professor would say, 'Well, Bill, the rats we feed IPF to, can't get enough of it. They gobble it up, gorge it, won't eat anything else, tear each other apart fighting over it.' Whereupon Ron Dull, the cautious H & G Chairman, might say: 'But Kipper, doesn't that mean that IPF is *addictive*?'

What would happen then? Would Glop and Dull instruct H & G Organoleptics to lock up the Sunshine file and refer IPF to the government advisory Committee on Toxicity? Or would Glop say, 'Don't be dull, Ron. People want IPF. Our shareholders need it. It isn't addictive. Its *more-ish*', and market Hope & Glory 'Instant Passion Flavour Froo-Tees' with the jingle 'Passion is now and forever with Froo-Tees'?

Well, what do you think?

Meanwhile, people who starve and gorge themselves obsessively are told that they are suffering from behaviour disorders and referred to psychiatrists. And the flavours in our food remain a secret.

Whiter Than White

So flavours are a trade secret. What about other classes of food additives? Take bread. The wrapper of a loaf of sliced white bread in front of me includes a lot of prose about how the Health Education Council reckons that sliced bread is the best thing since sliced bread, and about the fibre, vitamins and minerals in white bread. The label omits to mention that white bread has lost most of the nourishment present in whole grain cereal.

In small letters, the ingredients list is: 'flour, water, salt, yeast, soya flour, vegetable fat, preservative E282, flour improvers E300, 920, 924, 926, 927.'

How many additives are there, then, in sliced white bread? There's calcium propionate (E282) which, in common with E280 and E281, delays mould growth, and may be used in quantities up to three parts per 1,000 of dough. There's ascorbic acid (E300) better known as vitamin C, used here as an antioxidant. There's L-cysteine hydrochloride (920), a stabiliser. These three additives are not thought to present any health problem (vitamin C is a nutrient, of course); although propionates are reported to cause headaches and skin rashes among bakery workers handling these chemicals.

'Improver' sounds encouraging, but the word is used for a miscellany of additives that improve flour and bread for the miller and baker but not necessarily for you. Potassium bromate (924) is a dough conditioner. At levels very much higher than used in bread it is very poisonous, but rat and mice studies carried out by the millers and bakers are officially now taken as good evidence that it is harmless in bread. Azodicarbonamide or ADA (927) is a dough conditioner, only recently permitted in bread, which is good news for the trade because, to quote a textbook, it produces 'a lighter, more voluminous loaf'. It is not poisonous, as far as anybody knows, but its function as an airpump is of benefit only to consumers who like a bite of fluff.

Chlorine dioxide (926) is a controversial additive. It is a gas used by millers to age or 'mature' and to bleach flour. Bleach is not used just to shift stubborn stains. Various bleaches have been used for many years in British bread. Originally the reason was to make British wheat blend with North American wheat which had become whiter naturally during transport to Britain by sea.

In 1927 a government advisory committee stated that: 'A staple and indispensable foodstuff such as flour, the purity and wholesomeness of which are of cardinal importance to the community, should be jealously guarded against unnecessary treatment with foreign substances', and therefore recommended that the three flour bleaches then in use, chlorine, nitrogen trichloride (also known as 'agene') and benzoyl peroxide, no longer be permitted.

This recommendation was ignored. In 1946 it was found that agene caused fits in dogs and other animals. As a result agene was replaced by another bleach, chlorine dioxide, in North America in 1949–50. The British millers got round to making the switch in 1955.

Most British flour used for bread is still bleached with benzoyl peroxide. It has no E number; indeed, it has no number at all. It is not declared on bread labels. Why not? The MAFF Food Standards

Division (Bread) could not tell me. But Roy Knight of the Flour Millers and Bakers' Research Association explained. As from 1986, by law, flour 'improvers' must be declared on bread labels. But benzoyl peroxide, which functions only as a bleach, is therefore not an 'improver', and need not be declared. So it isn't. So I asked Mr Knight: 'Might not some people who wanted to know about benzoyl peroxide in bread read the 1986 labels and be misled into thinking it wasn't used?' 'That's a fair point,' he said.

White bread in Europe is not bleached. In 1960 the Food Standards Committee asked the British embassy staff on the Continent to find out why and how foreigners 'managed without flour improvers'. Here are the answers:

Belgium – The prohibition in Belgium of 'mineral substances' and flour 'bleachers' probably arose out of public health considerations since they are referred to as 'harmful'.

France – The bleaching of flour was opposed by hygienists before 1912. Since then two criteria have governed the decisions of French ministers, namely, is the addition technically necessary and is it safe from the point of view of public health? The French Ministry of Agriculture does not consider that the bleaching of flour conforms with either of these two conditions.

Italy – The use of bleaches and 'improvers' has never been officially allowed, presumably on health grounds.

Germany – The German Research Council could not recognise any necessity for the bleaching of flour.

Are foreigners fuss-pots? British millers and bakers, anxious to export British bread to Europe, think so. But in 1974 the FSC remained anxious about 'improvers', and put chlorine dioxide, together with potassium bromate, on probation pending the results of animal studies whose results were published in a journal of toxicology five years later. These were accepted by government (without, as far as I know, any public mention of the fact) as sufficient evidence of safety.

If 'improved' rats and mice don't get cancer, does this mean that chemicals in white bread are safe for humans? There are three problems with 'improvers' and bleaches. First, they destroy most of the vitamin E in flour and may damage its essential fats content. They are, at the least, 'anti-nutrients'. Second, Dr Hugh Sinclair has suggested that the chemical reaction between bleaches and essential fats in white flour and therefore bread might contribute to the uniquely high rates of heart attacks in Britain. This is a highly speculative idea, but Dr Sinclair, the most distinguished living nutritionist in Britain, has a habit of being right. Third, it looks as if more and more people in Britain today are made ill by eating bread. There is evidence that some people are poisoned not by the cereal itself, but by chemicals in it. Bleaches are irritants, and benzoyl peroxide is a source of benzoic acid which, as used to preserve processed foods (in the form of E210) is now known to poison vulnerable people, particularly children.

So what, then, is the last word of government and its expert advisory committees, on the additives in flour and bread? Responsibility for official policy on food and health in Britain is split between two government departments, MAFF and the Department of Health and Social Security (DHSS). The central DHSS advisory committee on food and health is the Committee on Medical Aspects of Food Policy (known as COMA).

In 1981 a COMA report on 'Nutritional Aspects of Bread and Flour' was published. It represents the most recent scientific view of our daily bread. Its first recommendation was: 'The consumption of bread, whether it be white, brown or wholemeal, should be promoted and bread should replace some of the fat and sugar in the diet.'

This is why in 1986 sliced white bread carries seals of approval from the government-funded Health Education Council. But what about the additives in white bread? The first meeting of the COMA bread and flour committee was held on 6 June 1978 in Room D110 at Alexander Fleming House, headquarters of the DHSS. I have obtained a copy of the minutes of this meeting. Item 8.2 was a statement made by the committee Chairman, Sir Frank Young. It reads: 'The Chairman confirmed that the Panel's report would not be concerned with such additives as preservatives, bleaching agents, antioxidants or contaminants.'

The official line on Dr Sinclair's suggestion, and on all other concern expressed before and after the FSC report of 1974, on bread additives is: no comment.

The work of the Committee on Medical Aspects of Food Policy, the FSC, and all other government advisory committees on food and health is covered by the Official Secrets Act.

The Chemical Feast
Most books on additives to date have given people the impression that all additives in food are now identified with E numbers; or, at least, numbers. Wrong. Few additives in our food are identified on ingredients labels.

It isn't just flavours that are a mystery. Of everyday food that everybody eats, bread remains the best example. So, back to the wrapped sliced loaf. What are its secret ingredients? For a start, bread contains chemicals which are neither additives nor ingredients. The farming of grain in the late twentieth century is a highly technological business. Grain is sprayed with pesticides, herbicides and fungicides up to eight times in the field and in storage before it goes to the miller. 'Additive' means a chemical which is added to food by the manufacturer. Any chemical which is added before the manufacturer gets his hands (or rather his machines) on the food, is not defined as an 'additive'.

Chemicals designed to kill pests, plants or fungi, while highly toxic in concentration, are supposed to vanish before any food reaches the supermarket shelf. This supposition is untrue.

MAFF funds a laboratory at London Road, Slough, part of whose job is to check food for residues of chemical sprays in food. I have received a copy of a report, from the Proceedings of the British Crop Protection Conference 1981, from the Slough laboratory, on residues of organophosphorus pesticides in flour and bread. 'Flanders' milling quality wheat was purchased from a farm in Suffolk in 20 tonne batches. This wheat had, in the ordinary way, been stored for five months and sprayed with pesticides. It was then milled and baked. The results? Contrary to a generally held opinion that the breakdown of organophosphorus pesticides during the baking process ensures that acceptable levels in bread are not exceeded, these experiments with wholemeal flour showed that at least 40 per cent and often more than half of the pesticide survived the milling and baking process intact.

The maximum recommended levels of pesticide residues set by WHO were often exceeded. The levels of pesticides in white flour and therefore bread were also 'surprisingly high'. For example, 'approximately 50 per cent of the pirimiphos-methyl present in the white flour remained in the bread following experimental baking'. Perhaps, the report suggested, the maximum recommended levels were set too low. 'The levels of residues found probably do not represent a toxicological hazard but could lead to problems in grain or flour exported from the United Kingdom to other countries.'

In other words, the WHO guideline is set by a lot of foreign fuss-pots, and there's probably nothing to worry about, so let's spray and spray. But watch out, chaps! Those neurotic fuss-pots may put an analytical chemist on to Great British grain!

So that's at least six chemicals, for a kick-off, that may be present in your daily white or wholemeal bread: chlorpyrifos-methyl, etrimfos, fenitrothion, malathion, methacrifos, as well as pirimiphos-methyl. Yum yum!

Now we get on to additives which are 'processing aids'. Roughly speaking, these are either chemicals added to the machinery as distinct from the food or chemicals added to the food to speed up processing. This is a rather vague distinction. Indeed, I've yet to come across anybody in government or industry who can make real sense of it. After all, a 'releasing agent' (to stop products getting stuck) is still going to be present in the food on the supermarket shelf. But you won't find it mentioned in the ingredients lists, because it's used to grease the wheels, not the recipe.

Here is a list of the 'miscellaneous additives' allowed in British food, including bread: acid, anti-caking agent, anti-foaming agent, base, bulking aid, firming agent, flavour modifier, flour bleaching agent, flour improving agent, glazing agent, humectant, liquid freezant, packaging gas, propellant, release agent, sequestrant. Most of these are 'processing aids'. No mention on the label. Are any of them toxic? I have no idea. It's an Official Secret.

For a long time bread was exempt from labelling regulations. In 1979 the FSC objected to this exemption saying: 'It is no longer a defensible argument to propose to a consumer, who may be a busy housewife, that to discover the ingredients of, for example, bread or ice-cream she should visit the public library or buy the appropriate food standards regulation from HM Stationery Office.'

Food manufacturers have always resisted explicit food labelling. Their arguments include: terribly complicated, awfully expensive, 'housewife' doesn't want to know, all additives safe anyway, unfair on the small manufacturer, would increase the price of food. And so on. But in recent years, the Europeans, who are particular about what goes into their food, have forced the hand of MAFF and the manufacturers. That's why E for Europe numbers now appear on British food; the decision, nominally made in Britain, was in reality a consequence of Britain being in the Common Market.

Dr Lesley Yeomans of the Consumers' Association spoke on this point at a meeting of the Society of Chemical Industry held in London on 15 May 1985. She was asked by a chemist from Cadbury-Schweppes whether consumers now cared about additives. She said in reply: 'You fought for a long time to keep additives off labels. If you're suffering from a backlash from consumers, you only have yourselves to blame.'

But the manufacturers have lost a battle, not the war. Virtually all chemicals used in our food remain a trade secret. And that's official.

Ssch!

So bleaches are an Official Secret. But why is an Act we associate with spy trials and the dumping of nuclear waste used to cover deliberations, however sombre, on bleach, brown 'For Kippers', boil-in-the-bag soyburgers and all the rest of the British food supply?

Conservative MP Jonathan Aitken asked much the same question in Parliament on 16 May 1985. Knowledge of the safety and quality of food, he suggested, is 'a subject where maximum publicity is needed, not maximum security.' In due course a reply came from Mrs Peggy Fenner, junior Minister at MAFF. This is what she had to say: 'It is necessary from time to time for information on manufacturing processes or other commercially sensitive material to be placed before the members of these Committees in order that they can properly advise Ministers on matters before them. The signing of the Official Secrets Act by the members of these Committees is the way of ensuring that the integrity of this information is protected from unauthorised disclosure to commercial competitors.'

That is to say, a trade secret is an Official Secret. Why? What is the real meaning of Mrs Fenner's statement of government policy? As far as I can see, what she means, is that unless members of government advisory committees are covered by the Official Secrets Act, they will be liable to leak trade secrets.

Picture the scene. A middle-aged man hurries along the corridors of power at Whitehall Place, MAFF headquarters. A meeting of the FAC panel on 'Extruded Foods' has just ended. Committee member Dr F C F Blue (for it is he), visiting professor at the department of Rheology at Lymeswold University (known to his friends as 'Brilliant') takes the MAFF stairs two at a time, walks quickly out of the door and down the street to a telephone box. He puts a brown paper bag over his head and dials a direct 'hot' line.

'Hope & Glory, Glop here', says the voice at the other end. 'Bill, this is Brilliant. Kipper's smoked out the Sunshine boys, all right. I've got the Instant Bliss formula.' Excited exchanges follow. 'Mmm, uh-huh, yes, it's IPF, right enough. Based on ethyl 2, 4-hexadienoate as we thought. It's synthetic passion.' 'Brilliant, Brilliant' says Glop. 'I'll call it "Instant Passion". By the way, the Froo-Tee Foundation will be funding your Hope & Glory Laboratory for the next academic year. Squared the endowment away with your Vice-Chancellor? Good, good.' And Professor Blue hurries on down Whitehall.

Is this scenario (totally fictitious, of course) what MAFF fears, unless government advisors are covered by the Official Secrets Act? It would seem so. I asked Michael Jopling, Peggy Fenner's boss, Minister of Agriculture, Fisheries and Food, why his committees were officially secret. His reply to me, dated 13 January 1986, said: 'Committee members, who have access to the fullest and most detailed information not only on safety but also on composition and methods of manufacture, must respect the legitimate wishes of manufacturers not to release information that might be of direct commercial advantage to competitors.'

The Labyrinths of Power

In order to make sense of British national policy on food additives, and indeed on food and health as a whole, it's necessary to know some facts about government advisory committees. The recommendations of these committees do not bind government, of course. Indeed, troublesome recommendations are usually revised, ignored or over-turned. Nevertheless, expert committees are important, and mem-bers of these committees can be very influential – depending, of course, on what they say and how forcefully they put their views across. Besides, expert committees are the instrument of policy. Mr Jopling and Mrs Fenner do not pretend to be authorities on food additives. They listen to their civil servants, who, in turn, set up the committees and service them. Government will not agree that coal tar dyes, or bleaches, or any other food additives, are a hazard to health, unless so advised by a relevant expert committee report. That's the way it works.

The central MAFF advisory committee is the Food Advisory Committee (FAC) which was set up in November 1983, as an amalgamation of the rather more explicitly named Food Standards Committee (FSC), whose reports on flavours (1965) and bread additives (1960 and 1974) have been cited above, and the Food Additives and Contaminants Committee (FACC), whose 1979 re-port on food colours has also been cited.

MAFF committees always include people from industry. This fact is not concealed. The official statement as applied to FAC runs: 'The Chairman and members of the Committee are appointed in a personal capacity and have backgrounds in the food industry, the enforcement of food standards legislation, the medical profession and consumer affairs.'

In 1985 five of the fourteen members of FAC worked for the food industry: Dr William Elstow of Associated British Foods, which includes Allied Bakeries; Dr William Fulton of Unilever; Dr Tho-mas Gorsuch of Reckitt and Colman; Jasper Grinling of Grand Metropolitan; and Alan Turner of Cadbury Schweppes. Typically, government advisers from the food industry are heads of research or chief chemists for very large manufacturers of processed food. Dr Elstow retired from FAC in October 1985.

Another two members of FAC are governors of the food industry-funded British Nutrition Foundation (BNF). These are FAC Chairman Professor Frank Curtis of the Agriculture and Food Council Research Institute at Norwich, and Professor Ian Macdonald of Guy's Hospital (who in 1986 is BNF Chairman). Daphne Grose of the Consumers' Association is also a member of FAC. She confirmed to me that her work for government is covered by the Official Secrets Act. She resigned as a member of the BNF Information Committee in 1985.

Opinions of the British Nutrition Foundation vary. Dr Derek Shrimpton was BNF Director-General from 1982 to 1984, and resigned after differences of view with his colleagues. Interviewed for Granada TV's 'World in Action', transmitted on 7 October 1985, Dr Shrimpton was asked how he now sees the BNF. He said: 'Oh, as a front for the food industry and a tool of Whitehall'. This view is not shared by others. Professor Macdonald, for example, has said (during his BNF Annual Lecture, 1984): 'Fortunately, there is one body that, in my view, is ideally suited to meet the need to give unbiased nutritional advice and to state nutritional facts, and that is the British Nutrition Foundation.' The BNF certainly provides a forum for people from government, science and industry to discuss issues of mutual interest.

It is not suggested that Messrs Elstow, Fulton, Gorsuch, Grinling, Turner, Curtis and Macdonald, or any other members of any government advisory committee, have ever given anything but their best advice. That said, though, the views of those people who work as senior managers for industry are likely to be consistent with the interests and policies of industry.

Daphne Grose's colleague at the Consumers' Association, Dr Lesley Yeomans, was a member of the now defunct FACC, and has also served on another MAFF advisory committee, on Food Quality, whose report was issued in 1984. What has Dr Yeomans's experience taught her? On 15 May 1985 she addressed the Society of Chemical Industry on 'Food Additives: the Balance of Risks and Benefits'. She said at the meeting: 'The Ministry of Agriculture is first and foremost a department to protect industry. There's no doubt about this. There's no question in my mind but when it comes to decision-making, MAFF puts food industry interests first. But when it comes to food additives, does this really matter?'

Put in perspective, food additives are not dangerous, Dr Yeomans said. 'Man has got very effective defence mechanisms.' Her view is shared by Professor David Conning, who in 1985 was Director of the government and industry-funded BIBRA, part of whose job it is to feed additives to rats and see what happens. 'Man is endowed with a peculiarly potent immune system', he said in a paper published by the

BNF in 1984. He also said that while some additives might be a very minor cause of cancer, others – antioxidants – might protect against cancer. In 1985 Dr Conning became Director-General of the BNF. His colleague as Science Director of the BNF is Dr Richard Cottrell, also a toxicologist, who has also worked at BIBRA. Wags say that the BNF, like the British food supply, is all additives and no ingredients nowadays.

You may think that in order to make sense of official decisions about food additives, all you need to do is to get to grips with the workings of the FAC and its predecessors down the years. Sorry. It's not as simple as that. For a start, problem additives are also referred to the Committee on Toxicity of Chemicals in Food, Consumer Products and the Environment (COT), a DHSS committee, mostly made up of medical men.

But the key government committee on food, whose remit includes additives, is a little known body of men who report directly to the Cabinet Office. This is a sub-committee of ACARD (the Advisory Council for Applied Research and Development). Sorry about all these initials, but the corridors of power are indeed a labyrinth.

Professor Conning is a member of this ACARD committee, whose report on 'The Food Industry and Technology' was published in 1982. To compete internationally, the ACARD report said Britain must develop food processing technology, fast. Food irradiation is seen as an important development, as are foods made not on the farm but by biotechnology: 'The most significant applications of biotechnology in the food industry are likely to arise in food processing but may take a long time and require considerable expense to obtain safety clearance for foods and additives produced by a novel route. Even when approved they may not be acceptable to the consumer . . . The benefits to the consumer need to be carefully understood and presented.'

A new national Food Directorate is proposed to control British food Research and Development (R & D). This should include 'scientists working in the food industry'. Research centres should 'become a magnet to attract research contract funds directly from industry and a training ground for scientists and engineers, some of whom could be expected to enter the food industry where they would . . . be able to increase the rate of introduction of new technology.'

You have been warned. New food technology requires new food additives. Food R & D should be controlled in large part by the food manufacturing industry. That's the recommendation. As for the consumer, what is proposed is a PR campaign.

How might a national publicity campaign for food additives go?

How about television commercials, featuring the Coal Tar kids, Di and Stan, two lovable little ruffians who beat their spoons at the table at party time: 'E102, E102, just the stuff for me and you, E123, E123, such a treat for mummy and me. Yummy, yummy, yummy!'

And for an adult audience? A young wife wearing a Union Jack apron lifts the lid of a saucepan and gives an expert little sniff. Then with a loving secret smile she takes a test-tube out of her additive rack (which is where the old-fashioned herb and spice rack used to be). She opens the test-tube. A wisp of smoke curls round her face rather sexily. 'M-mm'. She tips the contents into the saucepan. Cut, to husband sitting down, tucking into his dinner. She turns to us, the viewers. We share her little secret. Voice over. 'Sulphamethoxynitrosamine. For the man in your life. Now in supper-size tubes.'

Or perhaps a mailshot 'putting additives into perspective', sent to every doctor in the country, discreetly supplemented by briefings of selected medical correspondents at the headquarters of Beast Public Affairs at 666 Sequestrant Street, London W1. The subsequent press release might run:

THE SECRET OF LIFE? Professor Anthony O Skidant of Elixir University, Mo, reported preliminary results of a double-blind cross-over randomised trial of 20,000 corpses over a three-score-years-and-ten period. At a seminar held in London yesterday, Professor Skidant reported preliminary results. 'Bodies are better preserved nowadays,' he said. 'In our view the reason is rising consumption of butylated hydroxyanisole and butylated hydroxytoluene, a common constituent of highly processed food. All over America embalming businesses are going into liquidation.' Does this mean that BHA and BHT prolong life? 'It's too early to say', said Dr Skidant. 'But Krunchy Krispettes may be a useful ancillary therapy, before or after death.'

The ACARD sub-committee that included Professor Conning was chaired by Dr Douglas Georgala of Unilever, who is also Chairman of the BNF Industrial Scientists' Committee. The ACARD report led to the formation of the Food Priorities Board, chaired by Sir Kenneth Durham of Unilever, that in December 1985 made further recommendations for the funding of R & D for food, including food additives. The Priorities Board is advised by the Food Quality Board whose 1984 committee included Dr Yeomans, and was chaired by Tony Good of Express Dairies. It is also advised by the Food Safety Committee, chaired by Professor John Norris of Cadbury Schweppes, whose report was issued in October 1985.

To be precise, the Food Safety 'Consultative' report was dated July 1985, was released by MAFF with a press notice dated 7 October, but was first received by the press on 14 October. Deadline for comment was 18 October. Asked why such a short deadline, a MAFF representative told me, 'It's a mystery'.

The report states that the safety of food additives is a subject of 'great importance' and is 'prominent as a cause for concern'. Its conclusion, though, is that 'there is no significant evidence of hazard associated with the presence of additives and contaminants in the United Kingdom food supply'. Accordingly, it recommended that research funding into non-allergic adverse reaction to additives be wound down. No research, no significant evidence.

Two months later the Food Priorities Board accepted government policy to cut public funding of research, and to increase funding from industry. This, said the Priorities Board report, could be done by a levy on industry: 'The spending of these levies must be controlled by representatives of those who have provided them. This could require the establishment of small groups representing the appropriate sectors of the industry who will be responsible for determining how the sector's R & D needs should be met.' In other words, government policy in 1986 is that Bill Glop and the boys from Hope & Glory Organoleptics will fund research into Froo-Tee toxicity and presumably also control use of the research findings. They would, of course, report any disturbing findings to the Committee on Toxicity. Naturally.

Up to a Point, Minister

How many colours are there, in the food you eat? It's no longer a secret. As from 1 July 1986 manufacturers are obliged by law to declare colours on the label. Take chocolate confectionery. The packet in front of me has this ingredient list: 'Milk chocolate, sugar, wheat flour, edible starch; colours E171, E122, E102, E110, E127, 133; flavourings; glazing agent (carnauba wax), antioxidant E320.'

So that's six colours: titanium dioxide (E171), a white colour; carmoisine (E122), a red coal tar dye banned in a number of countries including the USA; tartrazine (E102) a yellow coal tar dye; sunset yellow (E110), another yellow coal tar dye; erythrosine (E127), a red coal tar dye currently in 1986 under a toxicological cloud; and brilliant blue FCF (133), a red coal tar dye which is due to be accepted in Europe and given an E.

In 1979 a report of the government advisory FACC recommended that colours be banned by law in foods prepared for babies and young children. This recommendation was not accepted by government but, as already mentioned, manufacturers have now voluntarily withdrawn dyes and colours from such foods. This does not apply to confectionery, however heavily advertised on television during childrens' viewing hours.

Successive government advisory committees have not been enthusiastic about food colours. In 1924 a Department of Health report said: 'Colour is frequently used to cover up objectionable or inferior materials, or to give a factitious appearance, so that the articles so coloured masquerade for something which they are not.' In 1954 the FSC report on 'Colouring Matters' felt that dyed foods could amount to consumer fraud: 'We are aware that where the use of colour is such that the food could be challenged in the courts as being not of the nature, substance or quality demanded and the purchaser is thereby prejudiced, proceedings could be instituted under section 3 of the Food and Drugs Act.'

Are coal tar dyes a cause of human cancer? The FSC was sure the question has to be asked. 'The suitability for use in foods of "coal-tar" colours involves the question of acute and chronic toxicity and of carcinogenicity,' the report stated. Furthermore, 'Reports are accumulating of experiments which point to azo compounds and dyestuffs of the triphenylmethane and stilbene series as potent tumour-producing agents in experimental animals. The results obtained make it inevitable that the possible carcinogenic properties of coal-tar dyes cannot be ignored.'

Trade representatives pressed the Committee: coal tar dyes are nothing to worry about, was the line: there were no medical reports proving that anybody had suffered cancer or indeed other disease because of eating coal tar dyes. Besides, customers enjoyed coloured foods, confectionery in particular. FSC was unimpressed: 'We cannot accept the contention that, because "coal-tar" colours have been used in foods for many years without giving rise to complaint of illness, they are, therefore, harmless substances. Such negative evidence in our view merely illustrates that in the amounts customarily used in foods the colours are not acutely toxic but gives no certain indication of any possible chronic effects.'

So what's the story in 1986? MAFF and its advisors are anxious not just about the number of colours and dyes used in our food, but also about the sheer quantity used. My own estimate is that by the time they are twelve years of age, British children on average will have eaten half a pound of coal tar (azo) dyes. This total does not include caramel or other colours which are not azo dyes.

Here is what Michael Jopling, MAFF Minister, had to say, to a joint lunch of the Cake and Biscuit Alliance and the Cocoa, Chocolate and Confectionery Alliance in October 1985: 'There is no point in over-colouring. It would be sensible to see whether a reduction in the amount of colours used would still satisfy consumer demand.' Mr Jopling's last few words are a bit quaint. Britain's high streets are not full of shoppers with banners proclaiming WE WANT DYE or BRING BACK THE STAIN or SHOPPERS DEMAND BRILLIANT BLUE PEAS. But the meaning behind Mr Jopling's words is clear enough. What he is saying to the trade is: watch out, boys. I'm warning you. There's a problem. It's not going to go away. Go easy with the paint pots. Say 'when' a bit sooner to those chemicals.

And with colours, as with all chemical food additives, there's the issue of secrecy. That's not going to go away, either. After acknowledging that members of MAFF advisory committes are covered by the Official Secrets Acts, Mr Jopling went on to say, in his letter to me of 13 January 1986: 'But this is not the same as saying that important information on safety is deliberately withheld from the public. This is certainly not so. All toxicological references are published with the reports of the FAC.'

Well, up to a point, Minister. A new FAC report on colours and flavours is awaited in 1986. Meanwhile, the FACC 1979 report on colours is the latest official word. It is indeed furnished with references: 346 of them. Here are three: '123. Gaunt I F et al (1975). BIBRA Research Report no 14/1975. 124. Gellatly J B M (1968). Unpublished report to Unilever. [125–128 are four more unpublished reports to Unilever] 129. Goldblatt and Frodsham (1952). Unpublished information from ICI.'

Oh well! Say no more! Just check through Unilever and ICI's files, and Bob's your uncle! Mr Jopling is sensitive to this point. His letter continued: 'I have asked my officials to see whether there are ways of ensuring the greater availability of these references perhaps by lodging copies of such papers with the British Library, where of course they would be accessible by all.'

Picture the scene. A supermarket in Huddersfield. Jo, wife and mother of two, is buying food for the family. She parks the double pushchair and ruminates over the label of an instant snack meal. 'E123, hm. Wasn't there some dispute over the Andrianova findings?' she thinks to herself. 'Better check it out'. So it's a quick phone call to Dave her husband, and she's off on the inter-city to London. This is fun for the kids, and there's lots of scrummy food from the BR buffet car. Then it's a taxi to the British Library. She chains the pushchair to the railings, straps the kids in, gives them a wholemeal sandwich and a copy of Fd. Cosmet. Toxicol. to keep them quiet ('Long-term toxicity and carcinogenicity studies of cake made from chlorinated flour, "1: studies in rats" for Darren, and "2:

studies in mice" for young Timmy') and in she nips with her British Library ticket for a quick flip through the latest research findings of Professor 'Kipper' Brown and 'Brilliant' Blue and the team at the Froo-Tee Foundation, Lymeswold University. 'The current toxicology on amaranth, madam?' says the library attendant. 'Certainly'. And after a browse, 'Well, that's all right, then', she thinks to herself. And it's out of the library, unchain the pushchair, into the taxi and the inter-city, back to Huddersfield and the supermarket, just in time to make a nice instant snack dinner for Dave, Darren and Timmy.

In reality, this is all beyond a joke. The truth of the matter is that every single expert advisory committee since the 1920s that has looked into the use of colours and dyes in food has been worried – seriously worried. Governments all over the world are kicking colours and dyes and other cosmetic and harmful additives out of their national food supply. This must happen here and now in Britain. It is time for our elected representatives in Parliament to realise – and many of them are realising – that commitment to a healthy British food supply is a vote-winner. And that means war on the additives that adulterate and contaminate our food.

5 Children At Risk

FELICITY LAWRENCE

Children are particularly vulnerable to harmful additives, and yet they tend to be precisely the people who consume them in high quantities. White bread, jam and children's foods such as fish fingers tend to be high in additives. Moreover, children consume large doses of chemicals from snacks, soft drinks and fun foods. Some of the damage done has already come to light in medical studies. The long-term effects are unknown. Children are being made the guinea pigs for a huge experiment whose results may only be discovered when it is too late.

What would you feed a young child? Fish fingers, sausages and packet mash, an instant pudding? A pack of sweets or crisps as a reward, a biscuit and a soft drink in between meals, perhaps? Or a couple of quarter pounders of chemical food dye?

On the government's own estimates, the average child in this country will have consumed nearly half a pound of artificial colouring (not counting caramels) by the time he or she reaches his or her teens. That's almost 50 mg of dye a day.

The US government's Food and Drink Administration (FDA) has also produced figures. The average American child between the ages of one and five eats 59 mg of artificial colouring a day; between the ages of six and twelve, 76 mg a day. At the top end of the scale, children eating a lot of processed foods consume 121 mg a day up to the age of five, and 146 mg a day between the ages of six and twelve; the most extreme consumption of colourings was 312 mg a day.

But what about caramels? Add a can of cola a day, and your child could be consuming a further three-quarters of a pound of colouring a year. That's not counting the huge amounts of caramels in chocolate and confectionery, other snack foods, soups, cereals, cakes and biscuits. And that's just the colourings. We have no way of knowing how many additives – preservatives, antioxidants which accumulate in the body fat, flavourings and other unlabelled processing aids – our children are swallowing each day. But what we can be sure of is that they are consuming far more now than any previous generation.

Most of the foods which are popular with children, and with busy mothers, are laden with additives. We have come to think of fun food as food which is highly coloured and flavoured. We start them off young, with coloured and syrupy vitamin or 'health' drinks. We send them to school to fast food lunches, junk snacks from the tuck shop and weekend treats from the newsagent.

The younger the child exposed to additives, the greater the potential trouble for the adult. Children are particularly vulnerable to harmful additives in food. They do not have the same capacity for metabolising food as adults, and until their immune systems are mature, they are less able to cope with any toxic load. Because their bodies are smaller, any additives they consume are likely to have a greater effect.

The government body responsible for additives, the Food Advisory Committee (formerly the Food Additives and Contaminants Committee) has been considering additives in foods for babies and young children for several years. No report has been published. But meanwhile, government has banned several additives from these foods. Not because any of them are unsafe, you understand, but because they are not needed (though they are apparently 'needed' in foods for adults). The chemicals which are banned in foods 'specially prepared for babies and young children' are:

ANTIOXIDANTS

E310 propyl gallate

E311 octyl gallate

E312 dodecyl gallate

E320 butylated hydroxyanisole

E321 butylated hydroxytoluene

ethoxyquin

The gallates are common in nursery foods such as squashes, instant puddings, instant mash and cereals. The butylates are often found in such children's favourites as sweets, crisps, instant puddings, instant mash, mousses, biscuits and cakes.

PRESERVATIVES

E249 potassium nitrite

E250 sodium nitrite

E251 sodium nitrate

E252 potassium nitrate

Nitrates and nitrites are used in a wide range of meat products including sausages, ham and bacon, the sort of food that might well be given to children.

SWEETENERS
420 sorbitol
421 mannitol
acesulfame potassium
aspartame
hydrogenated glucose syrup
isomalt
saccharin
sodium saccharin
calcium saccharin
thaumatin
xylitol
These are common in soft drinks, jams and marmalades.
OTHERS
E460 microcrystalline cellulose
621 monosodium glutamate
627 guanosine 5'-disodium phosphate
631 inosine 5'-disodium phosphate
635 sodium 5'-ribonucleotide
2-aminoethanol
These are additives which are found in fish fingers, fish cakes, sausages, beefburgers, instant and tinned soup and crisps, to name but a few – all foods an average mum might feed to her children.

The problem is clear. Young children don't just eat foods 'specially prepared for babies and young children' – they eat a whole range of foods, and often precisely those sorts of foods which contain the most additives. Foods such as crisps, soft drinks and confectionery containing those banned additives are advertised in children's comics and on television during children's programmes.

Government reports acknowledge this. The 1979 FACC report on colourings talks of foods described as being either directly or 'by implication' specially prepared for babies and young children. But at what age a young child becomes man or woman enough to take whatever additives are pushed its way is not said.

The report on colourings recommended that use of added colourings in children's foods be prohibited. 'A case of need has not been demonstrated for the use of added colouring matter in these foods. We are not convinced that the presence of added colour is of any benefit to infants and young children or that it affects their acceptance of the foods, though it may well, of course, make them more attractive to those who buy them, and so have an influence on their choice.'

The food industry operates a voluntary restraint and does not put colouring in foods for babies and young children, according to the Ministry of Agriculture. But as the FACC report says: 'We understand in fact that the majority of such foods sold in the UK do not contain added colour, but we think that manufacturers' self-imposed ban should be given statutory effect.' The recommendation has been ignored.

The FACC review on modified starches in 1980, recommended that 'starches modified by the use of epichlorohydrin and/or propylene oxide should not be permitted in foods described either directly or by implication as being especially prepared for infants and young children, and the level of use of the other modified starches classified in Group B [i.e. fourteen modified starches with question marks over their safety] in such foods should be limited to not more than five per cent by weight.' This figure of five per cent was reached after 'consultations with industry' on what was technologically feasible. The recommendations have been ignored.

Despite the industry's 'self-imposed ban' on added colouring, children manage to eat even more colouring than adults. The colouring report calculated acceptable daily intakes – the highest dose at which an additive causes no obvious toxic effects in animals divided by a factor of around 100 for good luck – and compared these with what it thought the average child and the average adult would eat. For tartrazine (E102), it estimated that the average adult eats 4.6 mg a day. The average child eats 6.7 mg a day. Amaranth (E123): the average adult consumes 1.4 mg a day; the average child 1.8 mg. Chocolate brown HT (155): the average adult 9.8 mg; the average child 14.6 mg. Yellow 2G (107): adult 1.38 mg; child 1.39 mg a day.

The safe dose of yellow 2G for children was calculated to be 0.25 mg a day. The average child is therefore consuming over five times the 'safe dose' of this colouring. The average adult consumes twice the 'acceptable daily intake'. What does the report suggest should be done about yellow 2G? Nothing. The committee simply put yellow 2G on the B group of colourings (seventeen in all) whose safety is in doubt.

What is the effect of a lifetime's exposure to repeated doses of additives from an early age? No one really knows. General practitioners have already noted an increase in the numbers of people who seem to be sensitive to additives (see chapter 6). The incidence of childhood diseases is rising in Britain. The rates of adult diabetes, eczema and asthma, for instance, have increased between two-fold and six-fold in twenty-five years. And work at the Great Ormond Street Hospital for Sick Children and at the Institute of Child Health has recently highlighted children at risk. There Professor John Soothill and Dr Joseph Egger have been conducting double-blind controlled trials – the sort medical experts prefer – which confirm what many parents have suspected for years.

The first of these looked at children who were suffering from migraine, and the results were published in *The Lancet* in 1983. The trial involved eighty-eight children, aged from three to sixteen, who had suffered from headaches at least once a week for six months or more. Most of the children also suffered from other ailments, including diarrhoea, abdominal pain, aches in their limbs, fits including epileptic fits, rhinitis, recurrent mouth ulcers, vaginal discharge, asthma and eczema.

The researchers then tested the children for allergies to various foods and to food additives. To test their reaction to artificial colouring and preservative they were given tartrazine (E102), sunset yellow FCF (E110), and benzoic acid (E210), in the form of a dose of orange squash. On the special diets, seventy-eight of the children recovered completely, and a further four improved greatly. Only six did not improve at all. All but eight relapsed when they went back on to their normal diets. Many of the other ailments cleared up at the same time.

Additives were not the only things to cause migraine in the children – cow's milk affected several children – but benzoic acid and tartrazine were both among the commonest provoking substances. Children who reacted to tartrazine did not necessarily react to benzoic acid, and vice versa. The authors also noted: 'Processing a food may affect its tendency to provoke symptoms; some patients reacted to white wheat flour but not to brown, and four reacted to bacon but not to pork.' With the exception of cigarette smoke and perfume, most of the things which had triggered migraines in the past, such as a knock on the head, no longer did when the children were on the special diet.

By 1985 Soothill and Egger had yet more exciting results. This time they had been looking at hyperactivity, a phenomenon which had attracted a lot of attention from the media, but which had until recently been dismissed by much of the medical establishment. Any suggestion that additives could be responsible for ill health and behavioural disorders among children was said to be wishful thinking by parents who could not face the fact that they had produced little monsters. A report produced in 1984 by the Royal College of Physicians and the food industry funded British Nutrition Foundation called 'Food Intolerance and Food Aversion', said of hyperactive children: 'It is all too easy to collude with parents, who cannot accept that psychosocial factors are to blame for their child's disruptive behaviour, by accepting that the child is suffering from food intolerance.'

The report also dismissed the significance of any intolerance to colourings as affecting only a tiny fraction of the population.

In Soothill and Egger's new study on hyperactivity at the Hospital for Sick Children, food colourings and preservatives emerged as the main offenders. Dr Egger was invited by the Hyperactive Children's Support Group to present his findings to their annual conference. The group has received over 100,000 requests for help from desperate families; the conference was packed with representatives – all parents from local groups.

Dr Egger began by describing the symptoms of the hyperactive child with the help of some slides. He (they are far more often male than female) may be fidgety, excitable, clumsy and impulsive. He is easily frustrated, unable to concentrate for any length of time, unable to sleep and, although he has a high IQ, he often does not do well at school. He may be alternately very depressed and very excited, or aggressive and prone to temper tantrums – at this point a slide revealed as evidence Egger's own bruised and battered arm photographed after a consultation with one hyperactive child.

Egger's study involved seventy-six children, aged from two to fifteen, who were unmanageable at home or at school, and 'all the children were socially handicapped by their behaviour'. Once again, many of them suffered from other ailments such as asthma, eczema, hayfever, headaches, rashes, aching limbs, persistent catarrh. The study showed that the additives tartrazine and benzoic acid provoked hyperactivity in four out of five of the children, although no child was sensitive to these alone. Other foods which produced allergic responses included dairy products and grains. Nevertheless, colours and preservatives provoked abnormal behaviour in more patients than any other substance. Again many of the other ailments also improved on the special diet.

The study concluded: 'The suggestion that diet may contribute to behaviour disorders in children must be taken seriously . . . Being on an acceptable diet did seem to make a remarkable difference to the lives of many of these families.' And on the subject of colourings and preservatives: 'They are particularly readily avoidable, since they have no nutritional value, and our findings strengthen the case for excluding them where possible from factory-processed foods and drugs, and for improved labelling.'

Egger told the conference that psychosocial factors are important in hyperactivity, but food and food additives are more important.

Parents are often made to feel guilty because they have produced a hyperactive child, and that is quite wrong, he said. There were roars of applause from the audience.

The idea that hyperactivity could be caused by food additives was first raised by a California allergy specialist, Dr Ben Feingold, in the 1970s. He had found that removing all artificial colourings and flavourings from the diet alleviated many complaints such as asthma, itchings, hives and skin rashes. It was only later that he found that these could affect behaviour too. He described many pathetic cases of children who had been given tranquillisers and sedatives for the best years of their lives in an attempt to control them and who had then responded to a diet without colourings, flavourings and salicylates (aspirin-like substances which occur naturally in certain foods, see p. 275). Industry sponsored studies to debunk his theories, and to some extent succeeded in discrediting him, but the Feingold Diet was already bringing relief to thousands of families. The Hyperactive Children's Support Group recommends a diet based on the Feingold theory.

Dr Jean Monro is a clinical ecologist who has spent ten years researching into immunology and biochemistry at the National Hospital in London. She has also successfully treated many cases of allergy by removing chemicals from the diet. Speaking at the same conference, she said: 'The incidence of allergies is rising. The incidence of asthma, eczema, hayfever and hyperactivity has increased . . . six per cent of the population now suffers from asthma compared with four per cent after the war . . . there is no doubt that chemical pollution is responsible for increasing illness in the community.'

The stories of hyperactive children – lost years at school, disrupted family life – are often tragic. These are the extreme cases which can be uncovered, once we have opened our eyes to the problem. But what of the less dramatic, but perhaps equally damaging, effects of additives in the long term? As Dr Vicky Rippere, a clinical psychologist at the Maudsley Hospital says: 'Over the past twenty-five years the food industry has carried out a gigantic experiment which would never have been allowed in laboratory conditions.'

Who knows what the results will be? It is our children who are bearing the burden of risk, for they are the unwitting guinea pigs.

6 The Size of the Problem

PETER MANSFIELD

Officially, then, all additives allowed in food are safe. Health problems caused by the chemicals in our food are rare . . . or are they? Many general practitioners feel that additives are a much more common cause of ill health, although some are reluctant to face the truth.

Changing Times

In midsummer 1985 the Department of Health and Social Security made a small but significant gesture. Professor Donald Acheson, the chief medical officer to the Minister of Health, wrote to all doctors to introduce a new information leaflet called 'Food Additives: Identification by Serial Number', published by the Ministry of Agriculture, Fisheries and Food. His letter and the leaflet appeared nicely in time to help doctors cope a little better with a wave of new public interest in everything to do with additives in food.

The letter accompanying the leaflet read a little differently from previous official statements, which have tended to dismiss as very small (a fraction of one per cent) the proportion of people affected by allergies to food, never mind to food additives.

Dr Acheson has not previously pronounced on the subject so far as I know, which gave him the unusual advantage for officials, of having no words to eat. So he was able to admit that 'some individuals may experience allergic or other adverse reactions to food additives . . . the incidence of such reactions is not known with certainty, nor is it possible to be sure which additives are most likely to cause reactions.' Then back to the old line, 'but it is generally accepted that the frequency of reactions to food additives is considerably less than that of reactions to natural food. Then another concession: 'However, while specific foods can generally be easily recognised, the identity of additives used in food is usually not obvious.'

Medical opinion is evidently shifting, and is a little less confident than it was. But there is still quite a gulf between official opinion and the actual experience reported by doctors who are prepared to acknowledge what they see.

General practitioners are usually the first people to be asked for help, and so they are the ones best placed to recognise a new phenomenon. Take the area I am most familiar with. Scarcely a week passes without contact with more new people with puzzling symptoms. I have seen a good many of these symptoms clear up with a change of diet. In many cases a change of food may be involved; but in many others an improvement is seen when the patient simply avoids additives.

Straw Poll

From the many dozens of patients who have sought advice I have chosen five cases for their simplicity and variety, not as strict representations of the situation in the area but as illustrations of the sort of problems that can be expected.

To obscure their identity from friends and relatives I have disguised the cases and renamed the individuals; but I have probably kept close enough to the letter of the truth for them to recognise themselves.

Andrew

The most recent case was Andrew, a young lad of fifteen who has suffered from asthma since his tonsils were removed at the age of four. By his fifteenth year, he was producing copious phlegm from his windpipe for seven days each month. Medication had disappointing results. To cap it all, he began to suffer from migraine during the same year, for up to three days at a time every week or two. The pain was intense and disabling, and frequently made him sick. No remedy worked very well, but the intensity of the headache was reduced by a pain-killer.

He already knew when he came for advice at the end of October 1985 that he was sensitive to cats, dogs, house-mites and feathers, which he cannot avoid altogether. So emphasis went first on other stresses or toxic burdens he could change, like his diet. Reduce the toxic load overall, and he should be able to cope better with those stresses which are unavoidable.

His diet was fairly conventional and included manufactured breakfast cereals, refined flour, squashes, cakes and biscuits and meats. He was advised first to cut out sugar and a specified list of suspect food additives, and to eat only wholemeal flour products. He began taking supplements of nutritional trace minerals. Eight weeks later his migraine had ceased completely. He is now looking more deeply into the asthma problem, and is hopeful about it. He may well prove to be sensitive to some foods, and has begun a series of exclusion trials to check each food family for culprits. But foods were not the cause of his migraine – he still eats the same range of foods, but choses varieties without the irritant additives. It appears that it was these which were crippling him, for up to a third of the time, in what should have been the most vigorous period of his life.

Faith

Now we go back to 1979, and a small incident in an otherwise active and healthy life. Faith was then twenty-seven, managing a busy household as a farmer's wife. For about a week she itched intensely from a patchy red rash, which came and went freely and interfered with her sleep. Nothing very obvious was to blame, so she began checking over in her mind which seasonal foods she had eaten, in case one of these was responsible. Meanwhile anti-histamine tablets held matters in check and made sleep possible.

After a couple of weeks she had found the cause of the problem. While waiting for the main garden crop of new potatoes, she had used some packs of a brand of instant mashed potato. Whenever she had a portion, the rash began again and lasted a couple of days; as long as she kept off it, the rash stayed away. Being a thoughtful person she bought some fresh potatoes and tried those. To her surprise the rash did not come back; she had suspected that chemical residues from agriculture, present in potatoes bought in shops, were responsible. Instead she was forced to the conclusion that one of the additives included in the instant mash, or perhaps the residues of undeclared processing aids which were used by the manufacturers, had been the problem. She has avoided instant mash since, and the rash has not returned.

Sharon

Sharon is a vivacious and talented seven-year-old, tormented for years by severe eczema and asthma. She owes a lot to the intelligence and good sense of her parents, who have patiently worked through all the approaches they can to discover the cause of her complicated problems.

Many of the results were very confusing. But one was crystal clear, and has made her mother a tenacious customer for organically grown potatoes. Whenever Sharon eats shop potatoes her skin begins to get worse; yet on home grown or organically produced ones she is fine. Most brands of crisps upset her too, but she has found one make with a simple additive-free recipe that she can have occasionally without much trouble.

David

Before he was two years old, David had exhausted his parents' very considerable capacities for tolerance. He would not sleep, or could not. They had done everything right – breastfed him to a ripe age, loved him, given him lots of attention and fed him wholesomely on fresh produce. Whatever else was there left for them to do?

They were advised to supplement his diet with nutritional minerals, vitamins B_6 and C and essential fatty acids – all things it is still possible to be short of on the fresh produce produced by today's farming methods (see chapter 9). Nothing much happened.

So they tried two exclusion experiments, for a week each. Domestic chemicals made no difference, but he appeared to react when foods containing aspirin-like substances were resumed. These substances – salicylates – became famous when Dr Ben Feingold identified them as the cause of hyperactivity in many children (see p. 275). David's parents then excluded these from his diet, and his behaviour became much more pleasant. His problems are now reduced to a hypersensitive cough at night, which responds to treatment: they are working on its cause. When they get weary, they remind themselves how much worse he might have been if they had given him any of the foods containing additives such as azo dyes and gallates, which can provoke reactions in people who are sensitive to salicylates.

Peter

Peter's mother had regularly let him have all the usual junk food and additive-laden treats, such as squashes and burgers, that his four-year-old playmates had. After cutting out all the suspect additives for a week, they had a different child. From being erratic, violent, irritable and hard to please, he was transformed to a lovable, affectionate, creative human being. When he was also taken off dairy products, his catarrh cleared up almost completely. His mother's remaining anxieties arose from trying to piece together a wholesome diet out of the foods he can have. She took some time to gain confidence that Peter would thrive on the simple meals which her neighbours looked on with pity and incredulity.

That was a year ago. Now a few of those same neighbours have begun to follow her example and seen pleasing results in their own children. Peter's parents have not only secured his health, but have also learned for themselves how infectious healthy behaviour can be!

Stephen

Stories like Peter's are told by many parents of children around his age. Stephen's behaviour made his parents' first attempt to get advice a nightmare – they simply could not concentrate on what was said. But they were given a list of additives to exclude for a month before trying again, and they resolved not to take him with them the next time they went for a consultation.

But by then he was so reasonable that they gladly took him along to demonstrate the effect of simply removing colouring from his food. Most of his problem was caused by the artificial yellow dye in orange squash. On diluted pure fruit juice, he has, so far as I know, continued happy ever after.

For some, like Sharon and Andrew, the search goes on. For each one whose problem is solved by avoiding additives completely, I can find two who have trouble tolerating particular foods. But how do we know that undeclared chemicals are not to blame, even in these cases? (The problem of agricultural pollutants is considered in chapter 9.)

Controversy

These stories, and hundreds like them, defy professional opinion that food chemical intolerance affects only a small minority of people.

When governments seek advice, they go to highly reputable scientists and academics, people who have generally been cut off for decades from regular contact with the unselected suffering of humanity. By the time they have spotted a new trend, it may have been obvious to some of their colleagues working in the field, but few ordinary doctors enjoy high status within the medical profession or control the sort of research facilities to make their observations scientifically respectable. In its report published in 1985, the Food Safety Research Consultative Committee recommended to the Priorities Board that funding for research on food additives should be wound down. But without the funds for research there is little chance of providing incontrovertible evidence about food additives. We need the help of academics to weigh the general significance of experience being gained by particular practitioners and observers of local trends, if we are ever to establish a balanced perspective.

Meanwhile there is no consensus on the role of additives in ill health. Many doctors have the firm impression that allergies in general are much more common than they were twenty years ago, but there are no reliable figures on this because it is extremely difficult to determine which patient's symptoms are caused by allergy and which are not. Nevertheless, many individual cases, such as those described above, can be quoted in which the influence of food additives seems impressively direct. In the past, stories like these have simply not been looked for. Now that doctors are beginning to open their minds where will their discoveries end? All we have found so far are a few of the more obvious sufferers, and those most determined to get well. Many other sufferers could be undiagnosed. Even more difficult to determine is the question of how many people are suffering indirectly from additives, where chemical intolerance may aggravate other conditions they have or impair their ability to resist disease and ill health. The question is no longer whether chemicals can spoil life, but which ones and how much.

7   The Great Unknown

FELICITY LAWRENCE

*Many questions about the long-term effects of additives remain
unanswered. We do know that some additives cause cancer and birth
defects when fed to animals; that some cause mutations in laboratory tests,
and that some are linked with cancer and birth defects in people exposed to
them at work. How much of a hazard these pose to consumers has not yet
been established. But where there is any doubt, many experts feel the
interests of consumers should come first.*

'Human exposure to a weak carcinogen may need to be prolonged for
several decades before any positive effect can be detected, and no
assurance can be given that an effect will not be produced by a
lifetime of exposure to the unusually large amounts that are consumed in diet drinks by some children and young adults.' So wrote
Professor Sir Richard Doll and Dr Richard Peto, the former and
current directors of the Imperial Cancer Research Fund Cancer
Epidemiology Unit in Oxford in 1981. They were referring to
saccharin and other artificial sweeteners, but their comments high-
light one of the major anxieties about additives in food. Even leading
scientists cannot give clear answers on the links between additives
and cancer or birth defects and other long-term effects on our health.

Like other chemicals, additives can harm the human body in
several different ways. The problem is that testing may only detect
some of the dangers (see chapter 3).

The current system of safety testing will show up additives which
are acutely toxic, and these substances are not generally allowed in
food. However, some additives, which provoke allergic reactions only
in certain people, may well escape the safety net and be approved for
use in food. Now that doctors are beginning to acknowledge that a
problem does exist, we at least stand some chance of pinpointing
these additives as cause of the problem (see chapter 6). And although
it is small consolation to those who have suffered from allergic
reactions, any damage is usually reversible.

But additives may also affect us in other ways which are much more difficult to detect. They may cause mutations, genetic damage or birth defects. They may cause cancer, or interfere with the body's normal processes – the long-term effects are little understood.

For example, additives may affect the way the body absorbs and uses nutrients which are vital for it to function properly. The sulphites (E220–227) destroy vitamin B1, which is vital to the nervous system. They also affect the absorption of another B vitamin, folic acid, which is needed to produce genes and new cells. Many women in this country are short of folic acid because of poor diet. Additives can also lock on to nutrients – additives used as seques-trants, for example, may bind up with calcium and iron and prevent the body from using them.

There is also evidence that additives damage our health by interfering with cell metabolism and inhibiting the formation of new cells. They can disrupt the balance of essential chemicals within the cell and prevent other chemicals which the cell needs from being taken in. Antimicrobial preservatives, such as nitrates, nitrites (E249 – E252) and benzoates (E210–E219), are used to prevent the growth of bacteria which spoil food, but they also affect the natural balance of bacteria in the gut.

Additives may also disrupt the body's vital functions by interfering with enzymes. Every process in the human body depends on enzyme action – enzymes control the chemical reactions which make possible the digestion and metabolism of food, and the synthesis of materials to repair tissues, for example. So any interference with the function of enzymes can disturb the body's normal activity. Sulphur dioxide (E220), sodium nitrate and sodium nitrite as well as many food colours are enzyme inhibitors. Other additives, such as BHT (E321), interfere with the normal functioning of enzymes by stimu-lating overproduction. Too much of any one enzyme can throw the whole system out of balance by disrupting the usual chain of reactions.

If toxins are absorbed through the gastrointestinal tract, they are transported to the liver. The liver has innumerable functions – it not only metabolises and detoxifies chemicals, but also produces bile to digest fats, regulates blood sugar levels and stores vitamins. But with constant assault from toxins it can become overburdened and dam-aged. And if toxins get past the liver they may lodge in and cause damage to other organs.

Additives may also be mutagens – i.e. they can interact with the genetic material of cells and cause mutations. When this happens the regulation of cell growth and division is disrupted, and the change in cell behaviour can lead to the formation of tumours.

Cancer

The link between additives and cancer is little understood because cancer is a 'multifactorial' disease – in other words, most cases of cancer do not have a single cause but are related to a combination of physical and social factors interacting with an individual's susceptibility to the disease. The fact that cancer rates are related to social class and vary from country to country provides strong evidence that environmental factors contribute to the disease, and it is now generally accepted that many forms of cancer are preventable. But we have to recognise the environmental causes first. And that is not easy.

In the first place, there is a long latent period before cancer develops in humans – it may take twenty to thirty years for the disease to appear after the original exposure to the carcinogen (a substance which causes cancer). It is therefore very difficult to isolate a single cause.

Moreover, a substance which causes cancer in animals will not necessarily cause cancer in humans and vice versa, which means it is very difficult to establish from tests on animals whether an additive is carcinogenic or not. As Dr Erik Millstone of the University of Sussex and author of *Food Additives* explains: 'Only recently has there been any attempt to establish a quantitive estimate of the correlation between toxicity tests with laboratory animals and human toxicology for chemicals that are believed to cause cancer. This analysis, carried out by David Salsburg of Pfizer Central Research in the USA, suggests that the animal tests are successful in identifying carcinogens only some 37 per cent of the time.'

Given the uncertainties, many experts feel that consumers should be given the benefit of the doubt and that if an additive causes cancer in animals it should be assumed that it poses a risk to humans too. The International Agency for Research on Cancer, whose work is recognised by the US government, says in its 1982 report on chemicals and industries associated with cancer, that: 'It is reasonable for practical purposes to regard chemicals for which there is sufficient evidence of carcinogenicity in animals as if they presented a carcinogenic risk to humans.'

Once the chain of events which leads to cancer has been set off, the disease is often irreversible. Even if it were possible to disentangle the causes of the disease many years later, when symptoms are beginning to show, it would probably be too late. Treatments are themselves highly damaging and invasive, and anything which increases the risk of cancer at all is better avoided.

There is no simple relationship between the amount of exposure to a carcinogen and the development of the disease. In animal tests tumours can be caused by repeated small doses, but also by single ones. So the idea of a 'safe dose' for any substance which is carcinogenic is totally inappropriate. For this reason, legislation was introduced in the USA as long ago as 1958 in an attempt to protect consumers from the risks posed by additives which might cause cancer. A legal amendment called the Delaney Clause was introduced, which says: 'No additive shall be deemed safe if it is found to induce cancer when ingested by man or animal, or if it is found, after tests which are appropriate for the evaluations of the safety of food additives, to induce cancer in man or animal.'

The irrelevance of establishing limits for the consumption of additives linked with cancer has been recognised in the USA, yet safety testing and regulations in this country are still designed to devise a 'safe dose'. At the same time, the uncertainty as to what tests might be 'appropriate' to identifying a risk to humans (mentioned in the second part of the Delaney Clause) is frequently exploited by food manufacturers in arguing that they should be allowed to go on using even suspect additives.

The Nitrates Story

The case of the nitrates (E251 and E252) and nitrites (E249 and E250) highlights the problem of the 'safe dose'. Nitrates and nitrites are widely used to cure and preserve foods such as meat products, processed fish and cheeses. They are useful to the manufacturer because they give meat products a pinkish colour and add flavour, as well as inhibiting the growth of poisonous bacteria. But their safety has long been in doubt.

Nitrates can be broken down to nitrites, and nitrites in turn may combine with amines which are naturally present in food and in the stomach to form nitrosamines. Nitrosamines are known to be powerful human carcinogens, and analysis of cured foods has found them in cooked sausage, bacon and cured pork. The medical journal *The Lancet* said in an editorial in 1968 that the presence of nitrosamines in food was 'a matter of the gravest concern'.

But controversy has raged over the use of nitrates and nitrites as additives. Industry has pointed out that nitrites can also be derived from nitrates used as fertilisers, and that the levels found in vegetables and water may be far higher than those in foods with these additives. Arguing against restrictions on nitrates as additives, the industry-funded British Nutrition Foundation booklet called 'Why Additives', points out that if you ban nitrates in food, you ought logically ban them in fertilisers too.

Epidemiological evidence has added to the controversy. In a study published in 1985, Richard Doll and his co-workers found no relation between levels of nitrate and nitrite in saliva and the levels of gastric cancer in two different population groups. On the other hand in the USA, Professor Philip Hartman of Johns Hopkins University looked at various population groups and found a correlation between nitrate and nitrite intake and cancer. Epidemiological evidence on chemicals used at low levels is notoriously difficult to interpret – after all, it took epidemiologists nearly a quarter of a century to show that smoking causes cancer. Other studies have suggested that nitrites themselves (not just nitrosamines) are carcinogenic, but the evidence for this has been disputed.

In 1972 the Ministry of Agriculture's Food Additives and Contaminants Committee (FACC) recommended that the use of nitrates and nitrites be at least reduced. In 1978 it said: 'We reiterate our earlier recommendation that every effort should continue to be made to eliminate the use of nitrite as soon as practicable, and that every effort should be made to reduce the levels of nitrate.'

Despite the fact that the idea of a 'safe dose' is inappropriate in the case of carcinogens – it is known that a single exposure to nitrosamines can lead to tumours – industry has gradually managed to shift the emphasis, so that the debate now focuses on whether nitrosamines in the quantities found in food present a risk, not on whether they should be banned. Dr Fred Steward of the University of Aston, charts the additives' progress in *Cancer in Britain: The Politics of Prevention*: 'The FACC's review in 1972 referred to studies on the presence of nitrosamines in various foods, but observed that the amounts of nitrosamines so far detected were minute and far below the levels at which such substances had been biologically tested . . . the problem had been redefined by industry and government from a major issue requiring action to one of a long-term research programme to explore dose-response relationships at low levels.'

But even if the idea of a 'safe dose' is accepted, animal studies on the effects of low doses of nitrosamines are not particularly reassuring. Certain kinds of nitrosamine compounds increased the incidence of liver cancer in rats at doses as low as two parts per million. Industry experts themselves acknowledge that there is a risk. P Grasso, a toxicologist at the British Petroleum Group Occupational Health Centre says in *Toxic Hazards in Food*: 'The high incidence of tumours induced in rats at these relatively low experimental dosages makes it difficult not to entertain the possibility that a real hazard exists from the nitrosamines present in food.'

But once again the emphasis is on a 'safe dose'. He concludes: 'In theory, nitrosamines in food present a real carcinogenic hazard to man. They are potently carcinogenic in animals even at low doses and are present in many types of protein food preserved by the addition of nitrate. In practice, this hazard would appear to be very small indeed since the amounts taken in by man are several orders of magnitude lower than the levels studied in animals so far.'

The FACC recommendation to phase out the use of nitrates and nitrites in food has never been implemented, but MAFF was sufficiently worried in 1982 to introduce amendments to the regulations on preservatives. Sodium nitrate and sodium nitrite were banned in foods for babies and young children and the levels of sodium nitrate and nitrite allowed in cheese and cured meat were reduced. Uncooked bacon and ham, for example, may now only contain 500 parts per million of sodium nitrate and sodium nitrite, of which not more than 200 parts per million may be sodium nitrite. The US government is considering restricting levels of nitrites to 40 parts per million. The legal limit on nitrite in American bacon is already, at 120 parts per million, far lower than that allowed in Britain.

With so many doubts about the safety of the nitrates and nitrites, why has the government not taken notice of the advice of its expert committees? Industry argues that the benefits of nitrates and nitrites outweigh the risks, because they prevent the growth of the deadly bacterium which causes botulism. In fact, this argument only really applies to tinned meats and products which are not exposed to the air. With products which we cook, the risk of botulism is negligible, and refrigeration and good hygiene are effective alternatives for most products. The nitrates and nitrites were among the fifty-one additives listed by the supermarket chain Safeway when it announced that it would be phasing out 'unnecessary' chemicals from its own label brands.

As well as the nitrates and nitrites, at least four substances approved as additives have been linked to cancer in humans (see chapter 8), and at least thirty-three additives are suspected of causing cancer in animal studies (see charts pp. 110–236).

Genetic Damage and Birth Defects

Cancer is just one of the great unknowns. Many additives are suspected mutagens, and when mutations occur in reproductive cells, the effects are transmitted to successive generations. Additives may also be harmful to the foetus and cause reproductive problems which are not hereditary – damage done can cause abortions, miscarriages, still births and handicaps. Chemicals which are capable of causing this sort of harm are called teratogens, from the Greek word meaning 'monster'.

As with carcinogens, additives which cause mutations or birth defects are difficult to detect. The interval between eating an additive which is mutagenic and any visible effect may be several generations. Pregnant women are exposed to whole ranges of chemicals and it is rarely possible to isolate individual substances as causes of birth defects. We are forced to rely on laboratory tests – but substances which are mutagenic in bacteria or animals, and teratogenic in animals will not necessarily have the same effect on humans. And there may be additives which are mutagenic and teratogenic in humans but not in animals, so their effects are not revealed by tests.

Studies have already shown that various additives may be mutagenic and teratogenic. In 1970 Russian scientists reported that the colouring amaranth (E123) caused birth defects in rats and a reduction in fertility; later studies confirmed that amaranth was toxic to the foetus. Young rats and mice suffered severe brain damage after exposure to monosodium glutamate–MSG (621). The offspring of rats fed BHT (E321) were born without eyes in one study; in another, while apparently normal at birth, they subsequently exhibited abnormal behaviour when fed BHT during weaning. The significance of these studies has been disputed, but meanwhile amaranth and other additives which are under suspicion are still permitted in this country.

The FACC report on colourings in 1979 noted: 'The results of bacterial mutagenicity studies indicate that two of the constituents of brown FK are mutagenic. We recommend that the results of a further carcinogenicity study in the rat be made available within five years, together with the results of a teratology and mutagenicity study.' And of amaranth: 'Mutagenicity studies have shown inconsistent results . . . In view of the uncertainties, we require the results of a well conducted carcinogenicity study in the rat within five years.' No reports have yet been published.

Several additives in use in this country are suspected mutagens. They include the colours curcumin (E100), carmoisine (E122), amaranth (E123), red 2G (128), some types of caramels (E150), and brown FK (154); sulphur dioxide and the sulphites (E220–227), orthophenyl phenol (E231), hexamine (E239), the nitrites and nitrates (E249–252), saccharin, and the solvents ethyl acetate and isopropyl alcohol.

Additives which are suspected teratogens in animals are: amaranth (E123), sodium acetate (262), BHA and BHT (E320–321), carrageenan (E407), monosodium glutamate (621) and other gluta-

74

mates, saccharin, aspartame, caffeine and nitrous oxide.

The hazards from additives which are suspected carcinogens, mutagens and teratogens are little understood. Industry and government work on the assumption that a balance can be weighed between the risks and the benefits of using these additives. But for most consumers, few benefits are worth considering if they bring any increase in the risk of cancer, or the possibility of damage to future generations.

8 Danger, Additives At Work

MELANIE MILLER

Workers in the food and chemical industry are doubly exposed to the
hazards of chemicals in food: many have suffered as a result of their
occupational contact with additives. The damage done to their health
should serve as a warning of potential dangers to consumers, but the lessons
often go unlearned.

On the Production Line

Between 1984 and 1985, at a soft drinks plant in the Midlands,
people handling the powdered mixtures for making drinks developed
dermatitis (sore and blistered skin which is very susceptible to
infection). The rash started on their hands and spread up their arms,
even to the neck and face of one worker. The dermatitis was caused
by the mixture of flavours, sweeteners and approved food colours put
into the drinks. The workers were issued with gloves as a barrier to
the mixture of additives. But this aggravated the problem; the
clammy rubber made the pores of the skin open up, and rubbed in
even more irritant dust. One woman commented: 'I never had
dermatitis in my life; no skin problems at all. But as soon as I started
to work in the mixing area I got dermatitis on my arms and even on my
face. Dermatitis is bad enough, but I hate to think what those irritant
additives are doing to my lungs in the long term.'

In the research department of a food factory a woman developed
asthma and bronchitis. On two occasions she had such severe asthma
attacks that she would have died had she not received emergency
medical aid. It took her five years to find in 1984 that her asthma was
produced by tartrazine, a yellow azo dye (E102). Tartrazine is used in
all sorts of foods such as salad cream, lemon squash, smoked fish,
cakes, sweets, foods in batter and packet sauce mixes. She found that
eating the dye in food triggered the same reaction as handling it at
work.

In 1985 people working in a fruit processing plant in the south-east
suffered respiratory problems from exposure to sulphur dioxide
(E220), a preservative allowed in jam, fruit juices, beer and many

other foods. This is an irritating and corrosive gas, dangerous to the eyes. Even at low concentrations it affects the throat, chest and delicate membranes of the lungs, making workers at the plant very susceptible to bronchitis and more severe chest diseases.

In a factory producing seasoning mixtures for meat products such as sausages, workers handle large quantities of preservatives, emulsifiers, flavours and colours. A union safety representative describes what happens when phosphates are used: 'The phosphate is tipped from a bag into the mixture. It's a powder. The extractor fans are in the wrong place, so the air gets full of dust. Every time it gets dusty, people's noses start to bleed. It lasts for quite a while.'

Dave, aged twenty-seven, was working for a Tyneside firm of food dye manufacturers. 'After learning the process in the laboratory I was given a shed with nine large vats and told to produce. Suddenly after two months at the place, I lacked energy. I started to cough, I couldn't concentrate. I'd get these terrible sweats. My neck glands and stomach would swell up.' Dave had been working with 'benzine', a known cause of cancer, and with various derivatives of 'benzine'. By December he was in the Whittington Hospital; he had cancer of the lymph glands, Hodgkins' disase. Dave said he didn't know if any more of the men at the firm were affected like him. The turnover at the plant was tremendous.

These are just a few typical cases. Like consumers, food workers have to face cocktails of additives – they may be exposed to mixtures of up to fifty chemicals at any one time. Given the problems with animal testing of additives (see chapter 3), health problems among workers involved with additives ought to sound alarm bells for consumers. But how much can we learn from their experience?

Canaries in the Coal Mine

Unlike consumers, food workers are not only exposed to additives in the food they eat, they may swallow additives from the air in the factory, or carry it into their mouths from their hands and clothing, or they may cough up and swallow chemicals which have found their way into their lungs – and their intake of some additives by mouth alone is likely to exceed the recommended 'safe doses' given by government committees.

On top of that, food workers may breathe in additives or absorb them through the skin. Dust, fumes and splashing can all bring the chemicals into contact with workers' skin. Levels of exposure vary enormously but tend to be highest in sectors such as soft drinks, snacks, baking, confectionery and meat products. In general, the more processed the food, the higher the level of additives used.

Workers are likely to suffer most when they work in the areas where the food ingredients are mixed or in researching new food mixtures, but they may also be exposed if they work in packaging or even in offices attached to plants where dust and fumes are allowed to drift.

So, compared to consumers, people working with food additives are exposed to much higher levels of chemicals, and by additional routes. How relevant is their experience to consumers, and what sort of things can we learn about the safety of additives from the problems workers suffer?

For many additives there is a simple relationship between toxicity and the amount to which we are exposed. Food workers may take in large quantities orally, while consumers eat far smaller quantities. But where workers' health has been damaged, we have an early warning of potential hazards for consumers. Ill health suffered by workers can give clues about the hazards to consumers of eating small doses over a life-time, or can give a warning where consumers are eating considerable quantities of a particular additive, not maybe from one source but because it occurs in a wide range of foods, or because some people eat vast quantities of a limited range of foods. For example, some children eat lots of snacks and soft drinks that are laden with added colours, preservatives, sweeteners and flavours. Health problems among food workers handling salt and vinegar crisps should have been investigated for the sake of children eating large quantities of crisps as well as for the health of workers themselves.

Where an additive is irritant or corrosive to workers, it is likely to cause problems for consumers too. Workers may come into contact with the substance by several routes, but chemicals which are irritant can be so via all routes of exposure – the mouth, skin and lungs. The upper gut – the mouth and throat – is more vulnerable than the lower gut which is protected to some extent by mucous. Sulphites (E220 –E227), for example, can burn the skin and eyes, irritate the throat and produce coughs, asthma and even prevent breathing in food workers. Consumers eating or drinking them in large quantities, in fruit drinks for example, may find that they irritate the mouth and throat and provoke asthma attacks.

Many additives produce irritation of eyes, skin, nose, throat and lungs on contact. This can produce and aggravate a range of health problems including asthma, dermatitis and vulnerability to bronchitis. Additives that can irritate or corrode include acids, like benzoic (E210), acetic (E260), citric (E330) and sulphuric (513) acids, processing aids such as the hydroxides (524–6), and bleaches like potassium bromate (924), chlorine and chlorine dioxide (925–6).

There are at least sixty-nine occupational irritants among the 300 or so additives that have been formally approved in the UK.

For some sorts of ill health, the relationship between dose and effect is much more complex – and extremely low levels of exposure can cause enormous and irreversible damage. It only takes a very short or small exposure to some carcinogens to start off the process of cancer. We cannot say with any certainty that chemicals which produce cancer in animals will cause cancer in humans (although our knowledge is beginning to improve in this area). So the information we can gather from food workers' experience on additives as potential causes of cancer is all the more vital. In the same way damage to genes (which can be passed on to subsequent generations) may be caused by a tiny amount of a chemical – where workers have suffered obvious forms of this sort of damage, we should be alerted to the fact that consumers could be at risk too.

Cancer

It is only because cancer has been observed in workers exposed to certain substances that hazards to consumers have been shown up. Assessments by the International Agency for Research on Cancer have shown that carbon black (E153) has probably caused cancer in humans by skin contact, and possibly by inhalation too. Talc (553(b)), used to dust tins to stop foods sticking, has been linked with cancer of the lung and pleura. Mineral hydrocarbons (905), used to glaze dried fruit and sweets, and microcrystalline waxes (907), used to stiffen and polish some foods and as an ingredient in chewing gum, have been linked with cancer in humans too. Long after these problems were first recognised, the allowable levels of carcinogenic compounds in these additives were reduced. But it is debatable whether this offers complete protection for workers or consumers in the long term.

Some of the azo dyes that have now been banned as food additives were originally linked with cancer because of ill health noticed among people handling these dyes in the chemical and textile industries.

There are huge problems in finding links between additives and disease from epidemiological studies, because the full range of substances to which a consumer is exposed cannot be monitored, and it is almost impossible to find comparable groups of the population who have not been exposed to additives to make judgements. Nevertheless, it was through epidemiology that the links between smoking and cancer were first established, and monitoring cancer among food workers exposed to additives could give useful information.

Allergies

There are other sorts of ill health where the link between exposure and toxic effects is very complex. Allergic and other intolerant reactions are produced by chronic exposure or repeated dosing. Once a person has been made sensitive, single doses of minute amounts can trigger severe responses. Here again we can learn from workers' experiences. Common causes of reactions in food factories are mixtures of colourings and flavourings used for crisps, azo dyes such as tartrazine (E102), ponceau 4R (E124), brilliant blue (133), sunset yellow (E110), the natural colour annatto (E160(b)), the benzoates (E210–E219) and the antioxidants BHA and BHT (E320 and E321). All can cause asthma and dermatitis when workers are exposed. Individual food workers have found that an additive can trigger a reaction such as asthma when they handle it and when they eat it. The cause of allergic reactions in consumers can be very hard to pinpoint – it is only rational to make use of the information that is already available from workers. Food workers are allergic to many flavouring substances, and to complex cocktails of additives – here we have a warning of danger posed by additives which have not been systematically tested for safety, and we can ill afford to ignore it.

More than forty approved additives can provoke allergic or intolerant reactions among people occupationally exposed. Gum arabic (E414), karaya gum (416) and gum tragacanth (E413), all widely used as emulsifiers, stabilisers and thickeners in foods such as cakes, piccalilli, sauces and some cheeses, can cause asthma and occupational dermatitis. Some people become hypersensitive to gum arabic after breathing it in, so that tiny amounts of it can trigger reactions later, ranging from rashes to lumps and severe breathing difficulties.

Reproductive Hazards

Studies in Finland have found that women working in food processing have an increased risk of spontaneous abortion. Researchers also found that women in the food industry faced a higher than average risk of having children with musculoskeletal malformations and children who develop cancer before reaching the age of fifteen. Women in the baking sector were found to be especially at risk of the latter. Problems like this should be investigated thoroughly so that causes can be identified and removed. Some additives are known from laboratory tests to be mutagenic (capable of damaging genes). These cause irreversible damage even at low levels, and so pose a risk to both consumers and workers. (Additives suspected of being mutagenic are described in chapter 7.)

All Risk, No Benefit

Drugs are tested in human trials and when they are assessed government officials weigh up the 'risk benefit' equation – drugs may have adverse effects, but these may be considered acceptable if they help cure or control a much more damaging disease. With additives the risk benefit equation nearly always involves a risk for the consumer and the workers who handle it, and a benefit for the manufacturer. No one would try to justify human trials for new additives – but in practice food workers very often end up being human guinea pigs. It is doubly insulting to them that the evidence their suffering provides is often ignored.

There is no systematic monitoring of the health of food workers exposed to additives – either for their own sake or for the benefit of consumers. Recent cuts in government expenditure have meant that 80 per cent of the statistical data on ill health and accidents at work are no longer collected. The government committees responsible for advising on food safety (such as the Food Advisory Committee and the Committee on Toxicity) examine the safety of additives when they are eaten by consumers; they are not required to make recommendations for the safety of people exposed at work. The government body responsible for that is the Health and Safety Executive, but it tends to assume that additives present no significant hazard. Along with occupational hygienists, the executive assumes additives are safe because they have been approved for use in food, and are subject to a more complex control system than most other chemicals. And because ill health has never been properly monitored among those who work with additives, both authorities tend to assume there is no problem.

Even where no direct link between hazards to consumers and ill health in food workers can be drawn, it seems highly questionable whether additives which do such damage to those who produce our food are ones we really want. Do we need them? Are they ones which improve the quality of our food? And do they merit the sacrifice of workers' health?

Do we really need talc (553(b)), silicates and stearates (551–572) which can damage the lungs and increase the risk of a range of diseases including lung cancer in food workers exposed to their dust? Do we need potassium hydroxide (525) and calcium oxide (529) which can cause burns and permanent scarring to lungs and skin? Workers may be exposed to poisonous fumes – for example, calcium chloride (509, used to firm some tinned fruit and vegetables), and sulphuric acid (513) can react with other common substances to produce noxious gases.

Nitrates and nitrites (E249–E252) are not only linked with cancer and hazardous to consumers, but present further risks to workers – large amounts by mouth may produce nausea, vomiting, collapse and even coma. Repeated small doses cause low blood pressure, rapid pulse, headache, visual disturbances and sometimes feelings of depression and weakness. They are powerful oxidising agents and there is a risk that they might cause fires or explosions by reacting with other substances if they are not handled properly.

The amount of protection given to workers by their employers varies. Despite legal requirements, many firms have inadequate ventilation systems. In some factories workers are obliged to wear cumbersome personal protective clothing, rather than the additives being kept in closed systems. And there is generally little difference between reputable firms and others when it comes to worker safety. The examples of ill health listed above all occurred in reputable companies. Moreover, the problems are not confined to small sectors of the food industry. The use of nitrates, bleaches and propionates (mould inhibitors) is identified with meat processing, flour milling and baking. But the use of most additives is spread across all parts of the food industry, and so it is not possible to pinpoint particular sectors where workers are especially at risk. Women suffer particularly. The food processing and manufacturing industries employ more women than any other manufacturing industry in the country, and within it there is a strong division between 'men's' and 'women's' work. Women are concentrated in the so-called less skilled and dirtiest jobs, like production. The double shift of working in a factory during the day and in the home at night can combine to create undue stress, which makes them more susceptible to illness at work.

Some groups of additives pose particular dangers to workers because the very chemical characteristics which make them useful also make them potentially hazardous to humans. They include bleaching and maturing agents, antioxidants, preservatives, sequestrants, organic solvents, volatile flavours, strong acids and strong alkalis. Many of these are direct hazards to consumers too.

Removing the Offenders

Some good news to emerge from all this is that by campaigning, food workers have been able to limit or ban some of the additives which are most harmful not just to them but to ordinary consumers.

A branch of the Union of Shop, Distributors and Allied Workers (USDAW) recently reported that 'tartrazine (E102), a yellow colour, is used in certain products at the Biscuit Works. It is a potential health hazard to workers. As from the beginning of April, management has agreed to phase this chemical out over a period of time.'

At a plant making packet sauce mixes in East Anglia a few years ago the colouring amaranth (E123) was put into the dried mixture. A union representative who was on a safety course realised that amaranth was an azo dye and a probable health hazard – to food workers and consumers. Union representatives asked the managers at the plant to remove the dye and this was achieved with very little difficulty because there was already some concern expressed publicly about it, and its colouring effect was not essential in the sauce mix.

In 1984 people mixing a flavouring and colouring 'concentrate' for crisps were told that a brown dye, brown FK (154), was to be introduced into the mixture. A Transport and General union safety representative at the firm was concerned because he had heard that this colour might cause cancer in consumers. He sought out more information and found that brown FK is indeed a suspected carcinogen. Since it is safest to assume that there is no such thing as a safe level of exposure to carcinogens, the union members at the firm refused to handle the additive, and the management reluctantly withdrew it.

What's good for workers is often good for consumers too!

9 And That's Not All . . .

PETER MANSFIELD

To add to the toxic load from chemical flavourings, colourings,
preservatives and the thousands of other additives used in our food, we are
increasingly faced with contamination of our food supply from other sources
– fertilisers, pesticides, hormones and antibiotics used in farming all leave
their chemical residues in what we eat.

Most of this book is about the chemicals deliberately added to food
during processing. But that is not the whole story. Modern food is
produced in a chemical environment quite divorced from the image
of natural goodness we tend to associate with agriculture. Farming
today is an industrialised agribusiness heavily reliant on chemicals at
every step in the food chain. And the residues of those chemicals
inevitably contaminate the food we eat.

The Quiet Invasion
The saga of additives begins right at the beginning: with the soil. In
the early nineteenth century Justus von Liebig analysed the ash from
a sample of burnt soil and discovered that its principal ingredients
were nitrogen, phosphorus and potassium. And so the idea of
fertilising farm land with mixtures rich in nitrogen (N), phosphorus
(P) and potassium (K) came about. There is no doubt that on healthy
soil it can yield far larger crops. But this is at the expense of the
organic life of the soil, all of which had gone up in smoke in Liebig's
experiment. As yields increase, organic matter comes under strain
and begins to dwindle. As it declines, so do the crops, so yet more N,
P and K are needed to make up for it. This increases the strain on the
soil and on food plants, whose health deteriorates.

Agricultural crops from NPK-fertilised land become progressive-
ly weaker and more vulnerable to disease. At this point there are only
two possible solutions. Either you strengthen the soil, or you weaken
the pests. The former involves calling off the NPK fertilisers and
rebuilding the soil – a slow and arduous process, and very unprofit-
able for the manufacturers of NPK. So they understandably spon-

sored the alternative, and looked for ways of weakening the rest of nature to the same extent. And so began the story of pesticides.

This process did not really begin until after the second world war, because there were neither sufficient sources of the fertilisers nor the means to distribute them. But the development of nitrate processing chemical works was greatly accelerated during war time for explosives manufacture, so that by 1945 a massive and efficient industry was producing nitrates which the explosives industry no longer required. At the same time the petrochemical industry was expanding rapidly, and so the problem of supply and distribution was solved. The marketing men simply turned their attention from war to agriculture, plastics, machines, domestic chemicals and appliances. The 1950s therefore saw great material prosperity, and greater changes in agriculture, which have been sustained ever since.

Nitrate fertilisers now deliver 1.34 million tonnes of nitrogen to the land each year in Britain, a twenty-five fold increase in forty-five years. Over the same period the pesticide subdivision of the industry has grown up, and by 1984 it was distributing over 29,500 tonnes of insecticides, fungicides and herbicides a year (many of which have already been identified in tests as possible causes of cancer, birth and genetic defects and allergic reactions). The effort of distributing all this has also put massive quantities of burnt petrol and paraffin residues into the air, and the manufacture of metals for the machinery needed has created thousands of tonnes of fluoride waste for disposal.

That sketches in the background to the huge burden of chemical exposure that life is now subjected to. It begins in the soil, in mineral ores and oils. It then passes inexorably through plants, food animals, water and air to people.

Taking the Soil Apart

Liebig proclaimed the chemical composition of soil, and science has based its teaching on the chemistry ever since. His mistake, and ours, was to ignore the living matter which inhabits and manages the chemical mud.

The living things in the soil are much its most important feature. There are millions of them, in great variety – from bacteria moulds and fungi to rats and rabbits. They create the physical characteristics which ensure good moisture retention, free drainage, good aeration and ideal root conditions. They are extremely thrifty in their chemical soil-keeping too, binding essential nutrient minerals safely in store off season and releasing them for efficient root uptake just when required.

Natural crop rotation, composting and surface cultivation actively encourage all this, retaining soil fertility for future generations. There is practically no limit to the sophistication soil life can achieve if it is given the chance. Chemical fertilisation, intensive cropping and deep cultivation, on the other hand, break soil life down rapidly. The effects take a few years on the poorest land, maybe decades on rich soil, but they are devastating. Soil crumbs become dust, and lose their sponge-like behaviour with water. The ability to bind chemical nutrients declines, so that serious shortages and imbalances develop which make the crops increasingly dependent on regular dressing of fertilisers.

But these only make matters worse; if mineral solutions are sprayed on the soil, they not only remain unbound, but may fool the remaining soil micro-organisms into releasing from bond what little mineral they had in store. All the soil mineral salts are then in solution, easily overconsumed by the plants, and exceedingly vulnerable if rain should fall before the plants have had time to absorb them. When it does rain, as in Europe it tends to, the precious nutrients are washed away to drains and watercourses. The soil can actually end up with fewer mineral stores after fertilisation than it had before.

If the minerals in the soil are depleted, they can't be drawn up into the plant and contribute to human health. And while vital trace elements are lost, nitrogen, phosphorus and potassium from the NPK fertilisers become over-represented. Vegetables grown in this way contain worryingly high levels of nitrates.

Disturbance of nutrient balance is harmful enough; imagine the effect of pesticide sprays actually intended to kill the bacteria, moulds and insects that attack crops. Residues soak into the soil, and damage life below ground just as easily as above it. They soak into plant crops directly or get taken up through their roots, so that fruit and vegetables may be contaminated. A survey by the Association of Public Analysts found that one third of the fresh fruit and vegetables they sampled were contaminated with pesticide residues. One fifth contained pesticide residues at levels well over the limit at which they are supposed to be reported. But, of course, these will not appear on your food labels, indeed their presence is often denied by officials.

The Unknown Quantity

When you walk into the greengrocers, you have no way of knowing what particular chemical cocktail your purchases have been given. Tests will only tell half the story because contamination can vary. Take the case of the potato. The defoliant herbicides used to destroy its plant before the crop is harvested also filter through to contaminate the tubers underground, but those growing near the surface are much worse affected than others buried deeper. So potatoes from the same bag will be contaminated in grossly differing degrees, giving unpredictable and puzzling effects in people eating them.

Other problems arise from the impure nature of chemical fertilisers themselves. Most are contaminated with cadmium and other heavy metals which accumulate in and poison plants and animals. They are also contaminated with fluorides, which accumulate in the soil and therefore reach plant crops in increasing concentrations with repeated fertilisation. Animals can only cope with exposure to these in tiny amounts before serious health effects become apparent.

The overall effect of these changes is to disturb the balance of mineral nutrients in plants, and therefore in humans, and to contaminate the food chain with poisonous pesticides, metals and salts we cannot help consuming, and have difficulty getting rid of.

Fast Food on the Hoof

Chemical fertilisers beguile farmers because they deliver crops fast, and the same pressures are felt by those raising animals for food. Animals can be made to reach market weight so much sooner with a bit of chemical assistance and intensive feeding regimes that few farmers can resist the commercial pressure to use them.

New and intensive methods of rearing and feeding animals have left them more vulnerable to disease, and they often need medication for all sorts of ailments, particularly infections. Antibiotics are now given to most animals in low daily doses as a preventive measure, and they have the extra bonus of making the animals gain weight faster. The results of this practice are disastrous. We are breeding antibiotic resistant germs much faster than before, because we give them the chance to crack the antibiotic code at leisure, and they survive to breed the secret into new strains. If the fears of some bacteriologists are correct, the outbreaks of salmonella deaths in English hospitals and old people's homes, which have made the news in the past two years, are just the beginning.

Another trick in the growth promoter's repertoire is the use of steroid hormone injections or implants. These are highly controversial. Government argues that they are perfectly safe and that residues in meat are too small to do any damage; 'the meat contains hormones naturally anyway' is the usual argument. But ones which were said to be safe in the past have subsequently been banned. For example diethylstilboestrol, or DES, was found to produce genital cancers and malformations in children born to women who had been given it medicinally to prevent miscarriage. The levels of DES found in the blood of some animals sold for meat were similar to those which produced cancer in animal experiments.

As long ago as 1982 the EEC considered banning the two synthetic hormones, trenbolone acetate and zeranol, which are used in Britain today, because their safety was in doubt. As with food processing additives, Britain trails far behind other European countries in protecting consumers. The EEC finally reached agreement at the end of 1985 to ban the use of all hormones in meat within two years. Britain tried to block the ruling, and has secured a dispensation to use hormones for an extra twelve months.

There are no reliable figures on the extent of hormone implanting in Britain, because no records are kept. But the Institute for Research in Animal Diseases at Compton, Berkshire, estimates that between 30 and 50 per cent of meat for domestic consumption has been implanted. The absence of figures highlights the major area of concern. Even if it were true that correct use of implants should leave no harmful residues, there is no way of policing the practice. Control over a substance which persists in the blood for weeks after a single dose is very difficult. Implanting, though nominally under the direction of the vet, is left to farmers. They may double the dose, in the belief that twice as many implants have twice the effect, or they may do it accidentally, by buying an animal which has already been implanted. And there is already a flourishing black market in hormones – prosecutions of vets for supplying hormones for animals which aren't under their direction are reported regularly in the trade press. British farmers will not readily drop hormones once the ban comes into force and the meat inspectorate is too small to stop an already established black market flourishing. The problem is far from solved.

A Packet of Trouble

Before it ever reaches the food manufacturer, our food has already been exposed to several chemical pollutants. Then come the additives, and the packaging. Cans contribute traces of tin, cadmium and other heavy metals to their contents. Plastics give off gas from solvents used in their manufacture, and these can accumulate in drinks and oil in sufficient quantities to cause concern. Foil wrappings may contaminate their contents with undesirable quantities of aluminium and lead.

The latest thing in food technology is irradiation. The process can substantially increase the shelf life of fresh and packaged foods without the use of artificial preservatives. As consumers become increasingly concerned about additives in their food, industry is heralding this development as the new alternative, the greatest thing since sliced bread. As we went to press a DHSS report giving irradiation the go-ahead in this country was expected.

There is evidence, however, that food irradiation depletes the vitamin content of food. It may also increase the formation of harmful free radicals (p. 272) and so make use of more antioxidants necessary. And an irradiated diet has been shown to produce mutations in laboratory tests on animals. The evidence is disputed, but uncertainty remains.

Good Enough to Drink?

I had forgotten how good water can taste, until I drank some from a highland red deer moor. The flavour, as well as the refreshment was a simple joy, lost now to most of us.

Dense population growth since the industrial revolution has put great pressure on our traditional water supplies from rivers and streams, forcing us to drill for reserves deep in the earth. Some of these reserves have always contained poisons, and the effluents which have leached out of cities and the soil in recent times have spoilt the rest.

Let's begin with the oldest controversy, and the most instructive. In the history of marketing it must surely go down as one of the great triumphs. I refer, of course, to fluoride.

The Fluoride Fallacy

Some deep water springs have always borne a little fluoride; a few of the more contaminated ones caused obvious and well-recorded damage to the skeletons of people who used them regularly. For this reason, until the 1930s any water source with fluoride above one part per million (ppm) was condemned as unfit for human consumption. But over a narrow range of concentration, around 1.5 ppm, the only noticeable effect of consumption was a reduction to about half in the amount of tooth decay to be found in children.

This fact solved two dilemmas at once. In the late 1920s a chemist named Cox was funded by an American sugar manufacturer to find a way of reducing tooth decay without making people eat less sugar. At the same time, aluminium and steel manufacturers were regularly being sued by people who had accidentally consumed the fluoride effluents from their smelting plants (disposing of the fluorides in metal ores had long been a difficult problem).

Cox was the first to see that if these fluorides were deliberately dispersed in the fresh water supply of enough communities, at around 1.5 ppm, the metal manufacturers could get rid of all their fluoride stockpiles and the nation's teeth should rot more slowly. The scheme could even be sold as a benefit to community health. Brilliant!

Intensive information campaigns were directed at health authorities, doctors, dentists and professional associations, which resulted in many of them publishing their approval of fluoridation. The word 'toxic' was replaced by 'nutrient' in publications on the properties of fluoride. Yet there has never been any dispute that at levels as little as twice the recommended 'nutrient' concentration, toxic effects will be found in a large minority of consumers – starting with the mottling of the teeth. And there are many reputable scientific references to gastrointestinal effects, hypersensitivity, heart disease, cancer and a higher incidence of Downs' birth defects, to name but a few.

Nevertheless, in countries which produce or use a lot of steel the fluoridation lobby has successfully pointed to favourable professional opinion and schemes have gone ahead. Elsewhere they have been abandoned, or outlawed. In England dental caries has declined in any case, in all areas, whether fluoridated or not. The simplest and most effective way to prevent tooth decay is to avoid processed sugars; good diet and oral hygiene have always been the most potent dental resources, and the only safe ones.

The Nitrate Time-Bomb

Fluoride we add at the reservoir. Another problem starts years earlier, down on the farm. We left the story of nitrates (p. 86) as they were disappearing down land drains and soaking into the subsoil. It looks as if that happens to at least half of the nitrate fertiliser applied to the soil each year. Much of that appears immediately in surface water, where it accounts for substantial increases in the nitrate concentrations of rivers and streams over the last few decades. The rest sinks slowly into the subsoil.

This soaking process has gone on since chemical fertilisation began, and each year's dose has gradually been sinking deeper into the subsoil all that time. So successive layers of the earth's crust, like the rings of a tree trunk, record the fortunes of the corresponding year.

The deepest layers are now reaching the subterranean water courses which feed our deep wells and boreholes, so nitrate concentrations in ground water have begun to rise. They regularly exceed the WHO and EEC safety limit already, and are bound to get worse each year as the successively more concentrated 'rings' reach the water source. We are sitting on a time-bomb, which will go on exploding gradually over the next twenty years.

So the fifth of our 90 g daily intake of nitrates currently provided by water is going to increase, inevitably. That means we shall have to pay serious attention to the massive amounts that we get from vegetables – 70 per cent of the total. If we stop fertilising them heavily and let them grow more naturally, that intake could be cut drastically. Cut down further by not eating nitrates as additives in food (see p. 71) and we probably have the worst of the problem solved.

The Other Ingredients
Bacterially contaminated water would be lethal and smell foul, but the chlorine we use to make it safe is hardly an improvement. You can get rid of it by heating the water, but that would have killed off the original bacteria just as well.

Many old iron water pipes are now rusting, and to stop this staining the water some authorities now add polyphosphates. But these are chelating agents, which can enhance the absorption of heavy metals from the intestine. Since old lead pipes and new copper ones may, in soft water areas, contribute large amounts of heavy metals, there is a new hazard here which is worrying some scientists.

What's more, residues of more persistent pesticides are draining away with the nitrates and getting into water supplies in increasing concentrations. These are still very small, but it takes little of a highly reactive substance to provoke symptoms in susceptible people.

Taken together, the recipes now used for converting sparkling mountain streams into town mains water are unpalatable and hazardous. At least, as a last resort, we can clean up the water in our own houses to some extent by installing the domestic filtration equipment now being developed. And cheap water filters are available in many chemists, health food and wholefood shops.

Coughing Up
Air is not so easy, unless you fancy a gas mask! Direct personal measures to clean the air we breathe are not so practicable. Industries must be persuaded to clean the smoke before it leaves the chimney, and it is perfectly possible to clean up motor exhausts more efficiently and stop polluting urban air with asbestos from brake linings – we have only to follow the lead of other countries and insist that it be done.

But there is also action that we as individuals can take which would make a powerful difference to the air we breathe. We could easily cut down on domestic aerosol sprays and cosmetic solvents, losing nothing in personal convenience and saving money as well.

We still need to learn more about other factors. We drastically alter the electrical quality of the air in modern buildings and urban neighbourhoods, and we are beginning to discover the folly of this. Part of the tonic effect of seaside and mountain holidays has always been the electrical charge generated by moving water. We can put it back electronically, or we can stop taking it away by cutting down on the use of tube lighting, television screens, air conditioners and artificial pile carpets. Either way many people are already finding the effort worthwhile.

Controversy still surrounds the health effect of radio waves and electromagnetic fields. There are many things which are not yet fully understood – microwaves from ovens and radar, telecommunication transmissions, power cables, X-rays and cosmic rays.

Hope

But there is a ray of hope: many of these local phenomena can be changed, if and when we agree to it. In the meantime, it may be comforting to remember that life evolved against a background of more cosmic radioactivity than we experience now. This is no reason to increase it again, but enough to believe that concern for our environment has not come too late.

All the first steps are personal ones. We can choose to furnish our homes and cultivate our gardens with health in mind (through garden centres you have as much access to agricultural chemicals as farmers do). Growing vegetables on compost and in their natural season keeps their nitrate content low, which on its own makes organic gardening worthwhile. And we can patronise the organic farmers whose produce is increasingly available.

We can go easy on plastics, choosing glass for windows and containers. We can filter our water. We can avoid buying aluminium goods and this helps reduce the flow of fluoride waste from aluminium works. We can ionise our air, and save electricity to cut down on acid rain produced by power station emissions. We can find the courage to report spraying accidents on the farm, and express all our ecological misgivings to the authorities and representatives who can change things (especially at election times).

It all comes down in the end to the power of the purse. When millions of ordinary people change the way they spend their money, people in power sit up and take notice.

10   Popular Myths

FELICITY LAWRENCE

The debate about food additives has spawned a whole new mythology with characters ranging from Angela Average to Nutmeg and the Humble Spud.

In October 1985 a privileged group of journalists was given the opportunity to snatch a glimpse of the inner workings of the CIA. The Chemical Industries Association, as it is also known, presented the assembled group with its own views on food additives. The report it handed out, 'The Chemistry on Your Table', was judged by the medical journal *The Lancet* to be 'worthless'.

It contained several of today's more popular myths about additives. Many people in industry now privately admit that the battle has been lost, and that pressure for more responsible use of additives is too strong to resist. But if you were to stumble across a gathering of the industry's old warriors, huddled round their pints of propyl 4-hydroxybenzoate best, you might still catch some of the old stories. We record them here for posterity.

First an old favourite:

Everything that exists is a chemical.
All foods are chemicals.
Food additives are chemicals.
Therefore food additives are safe.

Or a jollier version of the lyric goes something like this:

You and I are chemicals.
Your best friends are chemicals.
Additives are chemicals.
Therefore additives are a girl's best friend.

This one is in the tradition of O-level chemistry, but apparently pre-biology and physics. The trouble is, chemicals removed from their biological and physical context behave very differently and can

93

have quite different effects on our health. Take sugars: in their natural form, bound up with fibre in sugar cane or beet, they do not rot the teeth or cause diabetes in those who eat them. But take them out of that context by refining them to the point where they become effectively pure chemicals, and the effects on our health can be devastating.

Approach the argument from the other side and you get a lament that goes something like this:

Natural foods are not things you can trust.
The Humble Spud contains poisons, naturally occurring.
More people are allergic to natural foods than to additives.
Therefore additives are things you can trust.

A central character in this epic is the Nutmeg. Now you could be forgiven for thinking that nutmegs were responsible for the degeneration of modern society, for they apparently contain hallucinogens similar to LSD, and there have been many cases of nutmeg poisoning.

But if some natural foods do contain poisons, this ought to be an argument for keeping any other toxins in food to an absolute minimum, to avoid overloading the system. What's more, potatoes, unlike additives, contribute useful nutrients to the diet. The benefits of eating them are likely to outweigh the risks. And it is extremely difficult to eat enough nutmeg to poison yourself; it starts to taste rather unpleasant in the concentrations required.

Some people certainly are allergic to natural foods, but how can we be sure in all cases that their allergies are caused by the foods themselves and not by undeclared additives or unknown chemical residues?

Eat enough of anything and you'll die. Men have been known to overdose on carrot juice.

It requires an extraordinary single-mindedness to overdose on carrot juice. Those who have achieved it, have drunk nothing but carrot juice for several days. The assumption is that you could not eat anything like enough of an additive to cause damage, without being equally eccentric. But government reports have shown that some groups of people can quite easily consume far more than the acceptable daily intake for particular additives (see p. 58).

Now enter Angela Average (the old warriors of the food industry always refer to the consumer as 'the housewife, she', and one source even talks about 'a gaggle of housewives'). Angela Average is generally portrayed as being none too bright, and needs to be told that:

Large-scale food processing is no more sinister than what goes on in the kitchen. Many kitchen processes are quite subtle experiments in physical chemistry.

Who knows what Angela is getting up to in her scullery? By the time she's finished beating the living daylights out of a battery egg for her mayonnaise, it's little more than an emulsion of broken droplets of oil hanging in a watery solution. Might just as well eat ready made and mixed chemicals. Well, not quite. The trouble is that, despite apparently going through all the same processes as the big manufacturer, Angela and the rest of us don't seem to be able to get the same results. When did you ever manage to produce the standard white sliced loaf with its rising damp?

Now we have the hymn in praise of our ancestors:

Man has been using additives like salt petre, none other than our old friend E251, to preserve his meat ever since he left the cave. Even the Romans knew about it. Colouring of food goes back to biblical times. Sulphur dioxide has been kicking around for hundreds of years. So, long live the additive.

What man hasn't been doing for very long at all, of course, is adding the huge numbers of chemicals to the range of foods and in the sort of quantities that manufacturers use today. The toxic burden which we face from all sources of chemical pollution is hundreds of times greater today than it ever has been in the past. The Romans were long into decline and fall before they thought of adding sulphur dioxide (E220) not just to wine but also to cereals, jams, marmalades, pickles, instant mashed potato, potato scallops, dried vegetables, dried fruit, beefburgers, sausages and sausage meat, tinned prawns, fresh seafood, packet soups, instant snack meals, pot rice, packet stuffings, yoghurts, fruit juices, salad dressings and milk shakes, to name but a few of the things it is used for today.

By this time our huddle of ageing heroes are well into their cups of foaming agent, and are beginning to have apocalyptic thoughts:

Convenience foods as we know them would disappear. No acetic acid and there would be no acid-based sauces, no mayonnaise; no antioxidants and bottled oil would not last. No sulphur dioxide and no dried milk, worse still no beer, no cider. No emulsifiers – no low fat products. No texture modifiers – no margarine. Bread would go stale. No thickeners and gelling agents – no gums, pastilles (no surely not that), no ice-cream . . .

Nevertheless, at the back of this book, on p. 243, you will find a list of products made without these additives.

Ah, but it would cost more.

This line actually does by way of a chorus as it pops up in several places.

However, the Consumers' Association recently sent a researcher to buy a shopping basket of fairly typical items from ten retail chains. The report commented: 'With the exception of jam, there seemed to be no evidence to support the suggestion that low additive foods cost more.'

But if it weren't for preservatives, we'd all be keeling over with botulism or fatal bacterial infections from food. Illnesses such as these pose far greater risks than additives.

The fact is that there are alternative methods of preserving most food which are widely available and effective – e.g. refrigeration, sterilisation, vacuum packing. But if products are handled badly during processing, or if inferior ingredients are used, preservatives may have to be added to counteract the effect of bacteria.

The next problem, it seems, is that the consumer can't read, and anyway she wouldn't want to have all the additives, including flavourings and processing aids, with their long names listed on her food labels. What's more:

There wouldn't be room on labels for small products like chocolate to mention all the additives used.

I have in front of me a small chocolate covered biscuit. The words on the packet include an ingredients list with the following: 'milk chocolate flavoured coating 50% (sugar, vegetable fat, butter, dried skimmed milk, cocoa mass, whey powder, orange oil, emulsifier: lecithin, flavouring), flour, sugar, animal and vegetable fat (including hydrogenated vegetable oil), glucose syrup, salt, orange oil, colours: paprika, caramel, emulsifier: lecithin, flavouring, antioxidant BHA.

But it's only the chemical names which make additives sound sinister. Azo dyes are, in fact, only synthetic organic colourings; doesn't that sound better?

This is the Windscale principle. Change the name to Sellafield, and hope no one remembers that there was some problem with it. American humourists Beard and McKie have proposed other new names for additives: abominine, odiose, noxides, detrimenthyl, disgustillates, malevolene, lethalicin, gum malefic, exacerbene, woebegene and phobic acid, but these don't seem to have caught on in this country yet.

Now we come down to questions of principle, such as the consumer's right to choose:

But they wouldn't buy foods with additives in them, if they didn't want them. And if a product doesn't look right, it doesn't taste right. Why shouldn't we have tinned green peas when fresh peas are out of season?

Yes well, Angela Average really has really let us in for it this time. When manufacturers took colouring out of processed peas and they went a nasty khaki colour, she objected and wouldn't buy them, thinking no doubt that something funny had happened to them in the processing.

Some tinned peas manage without artificial colouring, of course. But manufacturers argue that consumers expect their products to look or taste a certain way, and it is true that we have learned some pretty strange habits. Smoked haddock is naturally a dull off-white, yet we all seem to expect it to be bright yellow. Millions of pounds have been spent by food manufacturers over the years to build up these expectations through advertising – we have always been told that the best peas are the greenest and the best flour is whiter than white, and we can't always help believing them.

But who knows whether people buy these things because they really want them. The fact is that most people do not have a real choice when it comes to buying food with or without additives. Additive-free foods which are genuine alternatives to the products people want have never been widely available at the right price. When they are, we shall be able to tell what consumers really want.

Finally, the appeal to higher authority:

All these additives are allowed by government, and if they weren't safe, government wouldn't allow them.

This presumably is the tobacco principle. Tobacco is known to cause lung cancer, heart disease and emphysema, and can damage the foetus, but if it wasn't safe, government wouldn't allow it.

11   Plan of Action

FELICITY LAWRENCE and GEOFFREY CANNON

While certain sectors of the food industry stubbornly ignore any consumer pressure to remove unnecessary additives, other manufacturers and retailers are taking a lead. Concerned consumers, with the help of this book, can do a lot to help themselves when they buy their own food. But it takes effort, and not everyone has access to or can afford new additive-free brands. The time has come for government action if we are to improve the national food supply and the country's health.

The Time for Transformation

And now for some good news. Despite the lethargy of the government bodies responsible for controlling additives, despite strong resistance from those sectors of food industry which have most to lose, many manufacturers and retailers have realised that something has to be done about additives. The change of heart has been brought about by consumer pressure.

In June 1985 Safeway announced that it would be removing a list of fifty-one additives from its own label products over a two-year period. The head of public affairs at Safeway, Tony Combes, had a special reason to be interested. He is a coeliac and allergic to gluten. He had spent long years struggling against undiagnosed illness and has learned to be concerned about what he eats the hard way. Interviewed for *New Health* magazine he said: 'I know exactly what the parents of hyperactive children must feel. We're removing fifty-one unnecessary colourings, flavourings and preservatives from our own label foods and replacing them with natural substitutes. We're starting with the things children eat a lot of: fish fingers and orange juice.'

He went on to say that Safeway had decided that customers should be given a choice: 'They've still got the branded goods with all the colourings and preservatives in them if that's what they want. But people are buying more healthily now – the changes we are making are as a result of customer demand. We're not trying to influence the brand leaders, but time will tell. If customers don't buy their products either, they'll have to change them. What we really need, though, is for someone in government, preferably the Minister of Health, to develop food allergies, and then things will really begin to change.'

Safeway is not alone in realising that there is a marketing advantage to be gained from removing additives from its products. All the major retailers we approached in drawing up our list of alternative products (see p. 243), were keenly aware of the need to clean up the image of certain highly processed foods.

Sainsbury says it is reviewing all its own label products to see which additives are really essential. Because of adverse publicity and consumer pressure from groups such as the Hyperactive Children's Support Group, the company began by removing a few colourings from foods, particularly those liable to be consumed by children, such as fish fingers and orange squash. The antioxidants BHA and BHT (E320 and E321) have gone from many of the biscuits, and the emulsifier E472 from some. MSG (621) has been taken out of a long list of biscuits, snacks, cooked meats and convenience foods. Waitrose too have fixed upon the most notorious of the azo dyes and decided to remove these from some of its brands. British Home Stores are doing likewise: tartrazine (E102), monosodium glutamate (621), caramels (E150), brown FK (154) and some preservatives have been replaced in some of its products. At Tesco, Presto, Asda, Marks and Spencer, Boots, the story is much the same – additives policy under review, small but significant changes, getting rid of those chemicals which have hit the headlines and consumers are therefore most likely to be wary of.

Retailers tend to be on the sharp end; they are the ones who come into contact with the public, they have to deal with the recalcitrant customer who is making a fuss in the shop, and so it is hardly surprising that the lead has come from them. They also have enormous power: own label products account for about 30 per cent of all packaged grocery sales in this country. That means retailers can lean on manufacturers to give them a higher profit margin and so force them to cut down on real ingredients in order to win contracts, but equally it means they can persuade manufacturers to clean up their products if they feel consumer pressure is strong enough.

Retailers have not been alone in their moves on additives. Major manufacturers have also joined the bandwagon. For example, you can now buy wholemeal digestives with the following ingredients: rolled oats, wholemeal flour, sugar, sultanas, vegetable fat and hydrogenated vegetable fat, nibbed peanuts, roasted ground hazelnuts, sesame seeds, honey, wheat bran, raising agent (sodium bicarbonate). 'No artificial colours, flavours or preservatives', it says on the packet; nor any nature identical ones come to that. Best before three months are up.

An advertisement for a new range of pure fruit spreads told the story: 'No colouring. No flavouring. No preservative. No added sugar. No wonder it goes mouldy after three weeks. It's too good to last.'

There's no problem making a product without additives then; it just needs a thoughtful bit of promotion to re-educate consumers whose expectations have been built up by decades of previous advertising. Keep it in the fridge, don't expect shelf life to mean the time it takes for the English language to change sufficiently to make the product's name meaningless. This is real food.

What About the Rest?

Retailers and manufacturers have a dilemma, however. Where consumers have been vociferous against particular additives, they can ill afford to ignore the trend. But if they admit publicly that they are removing additives from some of their foods because they are harmful, what on earth do they say about all the other products? What will the consumer make of those instant snacks and packet mixes that have so few real ingredients they simply wouldn't exist without additives? A spokesman for the industry association the Food and Drink Federation (FDF), David Walker, admits, 'no one understands all that needs to be understood about additives'. In 1985 the FDF was worried. It produced a confidential memorandum about 'media activity on additives'. There was to be an unwritten agreement between retailers and manufacturers that, 'if a manufacturer determines to produce a product without the use of food additives he should not promote it by suggesting that the product is inherently safer, or the old one inherently more dangerous.'

Just as government says it has banned certain additives in foods for babies and young children, not because they are dangerous, but because babies and young children don't need them, so manufacturers only take additives out of products because the housewife has got a bee in her bonnet about them.

The Sainsbury line is that there are some additives, mostly colourings, 'where there is some information suggesting a small section of the population may be adversely affected by them.' Marks and Spencer points out that the additives it uses are, of course, all ones which have been approved by government, but the company goes on to hint at a breaking of ranks by saying that it always reviews any reports on additives in this country and abroad itself before allowing its suppliers to use them. Tesco argues that much of consumer suspicion about additives comes from a lack of understanding of the subject.

The general consensus among retailers and manufacturers seems to be that additives implicated in hypersensitive or hyperactive reactions are a safe bet for a PR job. If shoppers make a lot of noise about them, take them out, but don't rock the boat too much.

Nevertheless, consumers have made progress. Industry is rattled, and showing it.

The Food Additives Campaign Team

On 12 December 1985 a new action group, the Food Additives Campaign Team, or FACT for short, was launched at the House of Commons with support from MPs of all parties and from trade unions, consumer and health organisations.

The general policy of FACT is to bring about a transformation in the quality of the British food supply, by means of a much more responsible use of food additives, and in particular by seeing to it that consumers have much more say in what additives are allowed into our food.

Hearing of the launch of FACT, the FDF called a snap press conference at its headquarters. Interviewed later on Radio 4's Food Programme, Barrie Williams, FDF Deputy Director-General, said: 'We are cooperating with government and the medical profession to ensure that we can provide the right level of information, in terms of additives overall, to ensure that the *damage done to the individual in terms of any adverse effect that additives have can be rectified medically*' (our emphasis).

FACT is supported by a whole range of organisations, including consumer groups like the National Federation of Women's Institutes, Family Forum, Foresight and the Maternity Alliance; health groups and voluntary associations such as the Coronary Prevention Group, the London Food Commission, the McCarrison Society, the Vegetarian Society and the Soil Association; groups of health professionals such as the Health Visitors Association and the British Society for Allergic and Environmental Medicine; specialist action groups already well known for their concern about additives, such as the Hyperactive Children's Support Group, Action against Allergy, the National Society for Research into Allergy and the Asthma Research Council; and organisations ranging from the Campaign for Freedom of Information and Friends of the Earth to trades unions such as the Baker's Union and the General, Municipal and Boilermakers Trades Union.

At the launch of FACT, Jonathan Aitken, Conservative MP for Thanet South, and a long-time opponent of the abuse of the Official Secrets Act said: 'We know more about what goes into a pair of socks than what goes into a packet of sausages.' He argued that the consumer's right to information was more important than considerations of commercial confidentiality.

Three other MPs declared themselves as supporters of FACT, and were later interviewed for BBC 2's 'Food and Drink Programme' on 21 January 1986. Of the Official Secrets Act, Michael Meadowcroft, Liberal MP for Leeds West, said on that programme: 'It's bizarre that an Act that was designed to catch spies is now used to keep people from finding out what's in their food.' The Liberal Party has now officially adopted several points in FACT's manifesto and is committed to introducing tighter controls on additives.

MAFF had, in fact, sent a senior member of its staff to the House of Commons FACT launch. He had a message, he said. Everybody could rest assured. All additives are tested. All additives are safe. Asked by the meeting chairman to identify himself, he declined. (An Official Secret perhaps?) Besides, he said, 30 per cent of the population were allergic to something or other; why pick on additives? If that was typical of the view of MAFF, declared Barry Sheerman, Labour MP for Huddersfield, 'we have a government department guilty of criminal complacency'.

FACT and its supporters are convinced that food additives are a political issue, because British public health will improve only when we gain a healthy food supply. That means controls on the use of additives to contaminate and adulterate our food. And that, in turn, means government action.

Not many people know that Mrs Thatcher herself was once a food scientist. In 1950 she was working for J Lyons at the Cadby Hall factory in Hammersmith, West London. Her speciality was fat extension: she wrote a learned essay 'On the elasticity of ice-cream'. During her time at Cadby Hall she also devised fillings for Swiss rolls. So in 1986 we have a Prime Minister with a special understanding of food additives.

A New National Policy

What is needed is the reconstitution of the official government advisory committees which consider additives. No one who is employed or paid by firms in the food, drinks, drugs or agrichemical industries should serve on these committees. Unions, and trade associations such as the Food and Drink Federation, should, however, be eligible to delegate members. Members should also be delegated by a full range of relevant representative bodies such as consumer groups, medical organisations and food law enforcement officers. The meetings of these committees should be publicised, their hearings should be held in public and evidence submitted to them should be published together with the committee reports.

Research on additives needs to be independent of industry. It should be the responsibility of the Agriculture and Food Research Council and independent bodies including medical, consumer, worker and environmental groups. The results of tests should be published. The research should be paid for by the manufacturers who make additives, manufacturers who use them and by government. Enough money should be made available to finance full and adequate programmes of research, and work in this area should be increased, not wound down (as proposed by the government's Food Safety Research Consultative Committee report to the Priorities Board).

Food free from harmful additives, in particular staple food, should be universally available at reasonable prices. Decisions about the use of additives must be part of a general policy to improve the quality of the national food supply. And that means reducing the number of additives in food and banning additives which are known to be toxic. Other European countries manage quite happily with far fewer additives.

Where there is any doubt about the toxicological status of an additive, the benefit of the doubt should be given to the consumer and the additive should be withdrawn. Additives believed to be toxic, individually or in combination – as a result of animal studies, clinical trials, observations in the workplace, reports from physicians, or other epidemiological studies – should be withdrawn from food, cosmetics and medicines. All additives which have only been approved provisionally by government because safety data is inadequate or because they are a cause for concern should be withdrawn.

Permitted additives should be restricted to a 'positive' list of those believed to be safe. The list should include classes of additives not now subject to regulation such as flavours and modified starches. Permitted additives should only be allowed up to specified concentrations in specified classes of foods. A full list of such additives should be published in an easily readable form. Food which is liable to be eaten by babies and young children or advertised as such should not contain additives. 'Young child' should be defined as under five.

All additives, including processing aids and those used in the manufacture of ingredients for other products, should be declared on the label. So, for example, bread should list any additives used in the flour on its ingredients label.

The House of Commons Select Committee on Agriculture should, for a start, enquire into food additives. Its brief should include examination of the quality of the British food supply as affected by additives; the short- and long-term effects of additives on consumers and workers; the incidence of disorders in which additives and contaminants have a causal role; and international collaboration in the regulation of additives.

What You Can Do

Now you have a copy of *Additives: Your Complete Survival Guide* you personally can avoid additives you don't want to eat. Off you go to the supermarket, look at the small print on the labels, check the E and other numbers, compare them with the ingredients lists, look up the charts starting on p. 110 for more details and (if you haven't been collared for suspected shoplifting by this time) you can choose food identified as containing harmless additives, or no additives at all.

Get keen on the subject and bore your friends with your knowledge of the toxicology of acids, anti-caking agents, anti-foaming agents, bases, buffers, bulking aids, firming agents, flavour modifiers, flour bleaching agents, glazing agents, humectants, liquid freezants, packaging gases, propellants, release gases and sequestrants: and that's just a list of the classes of 'processing aids' allowed into your daily bread, and other processed food! You will need to learn some law, organic chemistry and food science and technology if you want to avoid all these: they are not declared on food labels. Even so, if you are prepared to take a lot of trouble, and if you live in a city or near a well-stocked supermarket, and if you buy all the food you eat, you can avoid almost all the additives in your own food. You may personally be all right, Jack – or Jill.

But not everybody has the freedom to make well-informed choices. What, for example, if you are poor, infirm or without a car and nowhere near a supermarket? Or what if you depend on other people for your meals? Your freedom of choice is limited if you are a baby, a child or a student, if you are in the armed forces, hospital or prison, or if you eat in cafés, canteens, pubs, restaurants or other people's houses. In all these situations you have no idea what has been added to the food you eat. In the real world many people are not able to make well-informed choices about the food and the additives in food most of the time.

Change will come only when the political policy makers realise that a food supply improved by more careful use of additives is a vote-winner. So what else can you do?

If you don't understand the labels on food, ask questions of the retailers and manufacturers. Be more discriminating about the foods you buy: refuse to eat poor quality products which are tarted up with suspect chemicals.

Write to the large supermarket chains and food manufacturers asking them when they intend to reduce the number of additives in their products. Ask them to supply additive-free brands of your favourite foods.

Write to your local MP expressing your concerns about additives. Ask him or her to raise the issue in Parliament. Write to your local papers, ask them to write articles about additives.

Raise the issue of food additives at meetings of your local council, school parent–staff association, trade union, community group or political organisation. Ask them to support the campaign for a more responsible use of additives.

Contact your local council or area health authority and press them to use their buying power to encourage manufacturers to produce better food with fewer additives – ask for fresher, better quality food in schools, hospitals, day centres, canteens and restaurants.

Organise a public meeting; start a local group. On p. 281 there is a list of organisations which can help.

Good Food For Better Health
In 1986 the Great British food supply is increasingly made up of saturated fats, processed starches and sugars and salt made to look like, smell like, taste like and feel like real food by sophisticated use of chemical additives.

Undoubtedly, we all in Western countries eat too much fatty meat and dairy products, and too much sugar and salt added at table. But any message that genuinely traditional British food should be avoided is misleading. What must be stopped is the technological degradation of the food supply by means of chemical additives; the entirely new and utterly untraditional process by which, apparently, a necessary ingredient in asparagus soup is 'brilliant blue FCF' (133).

People concerned about additives are no part of any war against the food industry as a whole. The two million people who work in the British food chain would be glad to make, distribute and sell good and wholesome food. Regularly, the farming weeklies carry stories of farmers protesting against the contamination of meat with hormones. The *Meat Trades Journal* has been full of the concern among the trade about wholesale unsafe use of dyes in meat products. Food manufacturers as a whole need the protection of food composition laws, requiring high standards, just as much as we, the consumers, have a right to expect wholesome food. A healthy food policy should be a crucial national priority. Premature deaths from heart attacks are being prevented in the USA, Australia and other Western countries. It's time to win the same war in Britain. A healthy food policy is good for the health of the nation. Good food means better health. It also means more jobs, a decent environment and better prospects for industry.

This book hopes to be part of a new public debate on food additives, involving health professionals, industry and politicians, and most importantly, you as a citizen.

PART TWO

How to Read a Label

Food labels will give you a certain amount of information, if you know how to decipher the codes. There are still several products which are exempt from the labelling regulations, but in general foods have to carry a list of ingredients. It's usually found in the small print on the side or back of a product, or in a fold of frozen plastic (food labels have a lot in common with street atlases).

Ingredients must be listed in descending order of weight, though manufacturers still don't have to tell you how much of any one ingredient a product contains, so you only have a very rough guide. If sugar appears first on the label, for example, it probably is the biggest single ingredient. But classes of ingredients do not have to be declared together. So in another product sugars, taken together, may form the principal ingredient, but on the label they may be split up under several names, for example, corn syrup, glucose syrup, dextrose, sucrose, maltose, fructose or lactose, and so not appear at the top of the list.

In among the ingredients you will also find some mention of the additives contained in the product. From July 1986 these have to be listed by a descriptive name, such as 'preservatives' or 'emulsifiers' and their official number, with an E code if they have been approved by the EEC, or by their full name.

Flavourings, modified starches and enzymes only have to be identified by a descriptive name.

What's in a Number?
Additives which are approved by the EEC have an E number assigned to them. Additives which have numbers but no E prefix have not been approved by the EEC, but are permitted in the UK. Several of these are being considered for approval by the EEC (see charts), but others may be banned in the rest of the European Community.

Many additives have no number at all, and full lists of them are not published. No one knows how many flavourings there are – conservative estimates put the figure at 3,500. Representatives from industry say 6,000. You will not be able to identify these individually on labels. Additives which are classed as processing aids also escape declaration. These include a wide range of chemicals from bleaches for flour to solvents for colourings and flavourings, additives which stop food sticking to machinery and so on.

The Ones That Got Away

In your efforts to avoid additives which are harmful, you will inevitably come across several types of food which have no ingredients label at all. These include some confectionery and chocolate, cheese, butter, most milk and cream products, fresh fruit and vegetables, wine and other alcoholic drinks, vinegar and carbonated water. There is no obligation to declare what goes into or on to these. Some manufacturers argue that they should not have to label these items because the list of ingredients would give away vital secrets to competitors or because the labels are too small. But others seem to manage quite happily – some producers here already label confectionery, while in the USA all confectionery has to be labelled and wine will soon have to be too. Check with our charts if you want to know what goes into these products.

What's in a Name?

The description on the front of the packet will also give you a clue as to the contents, but it may well be a cryptic one (it helps to be a *Times* crossword addict).

Flavoured vs Flavour

For example: a strawberry flavour yoghurt has not been within sight of a strawberry; a strawberry flavoured yoghurt has had a brief encounter with a strawberry or two; and a strawberry yoghurt has had a much more lasting affair with the real thing.

'Free from Artificial Flavourings, Colourings and Preservatives'

This may not be quite as reassuring as it first seems. Your understanding of the word 'artificial' and that of industry and officialdom are probably rather different.

Several flavourings are synthesised chemically but are called 'nature-identical', and are not counted as artificial if they have the same chemical structure as natural ones. But sharing a chemical structure with a natural product is unfortunately no guarantee that it acts in the same way (see charts on sugar). Flavourings are not regulated by permitted lists, and we have little idea how they affect us.

There are also several 'natural' colourings which are highly suspect. For example, caramels (E150) – burnt sugar in the ordinary cookbook – are called natural, even though most types are made with ammonia and/or sulphite these days. Their safety has long been in doubt and they get a warning rating in our charts.

And what about all the additives which are not flavourings, colourings or preservatives in products labelled like this? How safe are they? Manufacturers keen to get on the health bandwagon may splash 'free from artificial colourings, flavourings and preservatives' all over the front of their products, but make sure you read the small print too.

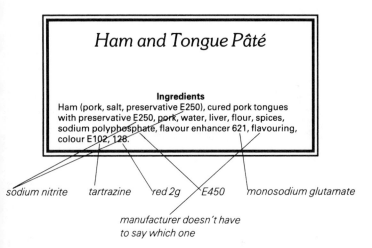

Ham and Tongue Pâté

Ingredients
Ham (pork, salt, preservative E250), cured pork tongues with preservative E250, pork, water, liver, flour, spices, sodium polyphosphate, flavour enhancer 621, flavouring, colour E102, 128.

sodium nitrite   tartrazine   red 2g   E450   monosodium glutamate

manufacturer doesn't have to say which one

Numerical Guide to Additives

Chart compiled by Dr Peter Mansfield, with Felicity Lawrence. Introductions by Felicity Lawrence. Research in supermarkets for food examples column, Adriana Luba. Thanks to Melanie Miller for her advice.

How to use These Tables

In the guide which follows, we have listed additives which are regulated in this country in numerical order – they have an E code if they have been approved by the EEC. When you see an E number on a food label, simply look it up to find its name, function and what is known about it. (If an additive is listed by its name on the label, use the alphabetical chart on page 229 to discover its number and/or position in this numerical chart.)

If you want to decide quickly how safe or dangerous any particular additive is, simply look at the rating given in the final column. (A rating also appears by each entry in the alphabetical listing.) Our ratings are unique, and take several factors into account – whether the additive is known to cause harm or not, and how serious the hazards are; whether research into its safety is adequate (if it isn't we feel strongly that the benefit of the doubt should be given to the consumers), whether tests have raised any doubts about it, the numbers of people it is likely to affect adversely; how widely and in what sort of concentrations it is used; and whether the additive is really necessary or whether a much safer alternative exists.

The ratings are as follows:

△ △ △   Avoid – unquestionably a hazard to many people

△ △   Beware – a definite hazard to specific groups of people and should be avoided by them

△   Caution – a possible hazard

√   Possibly safe, but controversial

√√   Safe – we can be confident this is safe for most people

$\checkmark\checkmark\checkmark$ Very safe

The EEC number coding groups classes of additives with similar functions together: so colourings mostly have numbers in the 100s; preservatives fall mostly in the 200s; antioxidants come between 300 and 321; emulsifiers, stabilisers and thickeners between 322 and 500; acids, bases and anti-caking agents make up the 500s; flavour enhancers are in the 600s; then there are no numbers until the glazing agents in the early 900s; and bleaches and flour improvers are 920 and on.

Our guide tells you how the additives are made, and why the manufacturers use them. (The glossary on p. 269 explains any terms that may be unfamiliar.)

We have compiled a list of the sorts of products in which each additive is likely to be found – and there may be a few surprises for you if you think you don't eat many additives. The list is the result of long hours spent scouring labels on supermarket shelves. We have also looked at the legislation which gives details of where additives may be used to fill in gaps. Additives which are used as processing aids do not have to be declared on labels, of course (see chapter 4). Our charts give you hints on where you are likely to find these.

The charts will also give you detailed information on what is known about each additive. Our 'restrictions' column enables you to see at a glance what governments here and abroad have said about any particular additive, and brings the information together in a quick reference form for the first time.

For the colourings, the 'restrictions' column will tell you whether any additive is approved or whether there are doubts about its safety in 1) the UK and 2) the EEC; 3) what the World Health Organisation/Food and Agriculture Organisation (WHO/FAO) has said about it; and 4) whether it has been banned as unsafe or unnecessary in any other countries.

For each of the other additives, the 'restrictions' column will tell you 1) whether the UK or EEC have put any limits on its use in recognition of health hazards and 2) what WHO/FAO has said about it. WHO/FAO has reviewed several additives and tried to establish maximum safe levels of consumption (called 'acceptable daily intakes' or ADIs) for them. These are not always reliable but do at least give some indication of safety. Where they have called for more tests, we have reason to be suspicious.

In the 'comments' column, we survey what evidence there is on each additive and its adverse effects. Where possible we have also given some indication of the safety margins. To do this we have looked at the recommended maximum daily intakes set by the EEC and/or WHO/FAO and compared them to estimates of normal daily consumption and the quantities in which they are permitted in foods. Where the safety margin is a factor of 100 or less, we have cause for concern. Safety tests are just not sufficiently accurate in our view to run that close (see chapter 3), and it would be easy to overconsume the additive. Where the margin is a factor of 10,000 or more, it is much more reassuring and scores a 'good'. We have drawn our information for this column from a wide range of respected sources to make the most detailed layman's guide yet published.

The 'risks' column is designed to be another quick reference, highlighting who is at risk, where particular groups are affected by any additive; and what the dangers are. In the case of cancer, the fact that an additive causes cancer in animals does not necessarily mean that it will cause cancer in humans. However, evidence that an additive is a carcinogen is a cause for serious concern, and we believe consumers should be given the benefit of any doubt. So we have included cancer in the 'risks' column.

At the end of the numerical guide, you will find a list of additives, most of which rarely appear on labels, or in books for that matter – they do not have numbers, but nevertheless find their way into our food. Sweeteners (including refined sugars) are covered under the section called flavour enhancers.

Flavourings, modified starches and enzymes are not governed by any specific regulations in this country. Even though they make up the vast majority of additives, there is no requirement to declare them on labels other than by a general description such as 'flavourings'. No complete list of these is available; even the Ministry of Agriculture says it doesn't have one. We have included general comments on these and their safety. For flavourings, see p. 226; for modified starches see p. 226; and for enzymes, see p. 227.

We have done our best, with the information currently available to us, to be fair, systematic and uniform in our judgements. Readers, whether consumers or from industry, are cordially invited to help us improve on the current state of knowledge by contributing further evidence (write c/o New Health magazine, Haymarket Publishing, 38–42 Hampton Road, Teddington, Middlesex, TW11, oJE) and it will be taken into account in subsequent editions. We would be particularly interested to hear from anyone who discovers food labels which fill in some of the gaps.

Key

△ △ △	Avoid – unquestionably a hazard to many people
△ △	Beware – a definite hazard to specific groups of people
△	Caution – a possible hazard
√	Possibly safe, but controversial
√√	Safe – we can be confident this is safe for most people
√√√	Very safe

FAC – the Food Advisory Committee – advises UK government on food policy and regulations, including additives

FACC – the Food Additives and Contaminants Committee, the predecessor of the FAC

WHO/FAO – The World Health Organisation and the Food and Agriculture Organisation of the UN which, among other things, sponsor a committee of scientists and representatives from governments and industry to review the safety of additives

MAFF – The Ministry of Agriculture, Fisheries and Food, which is responsible for additives in the UK

US FDA – The American Food and Drug Administration, part of the US Department of Health, set up to protect the public from unsafe food and drugs

mg – a milligram, a thousandth part of one gram

g – a gram, the basic metric standard of weight

kg – a kilogram, one thousand grams, approximately 2.2lbs

ppm – parts per million, the same as mg/kg. These are measures used to express permitted maximum concentrations of additives in food

ADI – (estimated) acceptable daily intakes – a measure used by the UK government, EEC and WHO/FAO to express their advice on safe limits of consumption for additives

Colourings (E100–E180)
Colourings are perhaps the least justifiable of all additives because they are purely cosmetic, and rarely add to the nutritional value of

foods. They are often used to disguise a lack of real ingredients or the use of poor quality foods in processed products.

The Food and Drugs Act requires government to restrict, so far as is practicable, the use of substances of no nutritional value as foods or ingredients of foods, but despite this, and the reservations of experts on government committees, the use of colourings is increasing steadily.

The colourings usually chosen by manufacturers are the azo dyes and those originally derived from coal tar dyes. These artificial dyes are more stable than natural ones, but many of them are known to cause ill health. Many of the azo dyes permitted here are banned abroad.

You will probably come across labels which say 'free from artificial colouring'; this is a complete red herring. It does not guarantee that a product is free from added colouring, nor that any so-called 'natural' colourings it contains are safe. Many 'natural' colourings, such as caramels (E150), are highly suspect.

Colourings have received considerable official attention. In 1979, the Ministry of Agriculture's Food Additives and Contaminants Committee (FACC) published a report on food colourings, which divided those in use in this country into various categories. Those that were fully approved went on to the A list; those whose safety had been called into question on to the B list; and those for which there was inadequate evidence to make any decisions about safety were classed as the E list.

The report put several colours on the B list, and called for further tests within five years (i.e. by 1984). It recommended that if satisfactory results were not published within that time, most of them should not be permitted in food. At the time of writing, no review has been published and all these additives are still permitted. The B list consists of E104 quinoline yellow, 107 yellow 2G, E120 cochineal, E122 carmoisine, E123 amaranth, E124 ponceau 4R, E127 erythrosine BS, 128 red 2G, E142 green S, 154 brown FK, 155 chocolate brown HT, E150 caramel, E151 black PN, E160(b) annatto, bixin and norbixin. Unbelievably, E131 patent blue V, which is on the E list, is also still in use, and unrestricted.

The FACC also recommended that colours be banned from foods for babies and young children, but the ban has never been implemented. Manufacturers exercise a 'voluntary restraint' and do not use colourings in babies' food as such, but young children do not unfortunately only eat foods labelled specifically as 'baby foods'.

The report also tried to establish 'acceptable daily intakes' (ADIs) for each colouring. These were based on the dose of a colouring which is toxic to animals, divided by a safety factor of 100 for good luck. The Committee then compared these figures to what they took to be average and extreme diets to see whether anyone was likely to exceed the 'safe' dose of any one colouring. It did not attempt to consider the effect of consuming several different dyes in combination – a situation which would be a much better reflection of real life. The normal intake of yellow 2G turned out to be more than five times the ADI. The Committee's conclusion: the ADI was probably too strict!

The WHO/FAO also monitors colours – its 8th report (1965) only accepted three for use in food – amaranth, sunset yellow FCF (E110) and tartrazine (E102) – and recommended that a further twelve be withdrawn. All of the three colours favoured in that report are now known to be highly suspect. The history of colourings is littered with such examples of dyes which were thought safe only to be found dangerous later. Only four of the synthetic dyes permitted in the UK today are also allowed by both the USA and the EEC.

Colour is important to the enjoyment of food, industry tells us. It is, which is why so many foods are such glorious colours in their natural unprocessed state. Our advice is to choose fresh food wherever possible and avoid cosmetic food colourings.

Colourings E100–E104

Code	Names	Source	Uses	Examples
E100	Curcumin, Turmeric	A root extract A root	Orange-yellow colouring	Breadcrumbs, banana yoghurt, sweets, dried savoury rice, froze pastry
E101	Riboflavin, Lactoflavin, Vitamin B_2	Occurs naturally in many foods. Prepared from yeast or synthesised chemically	Orange-yellow colouring	
101a	Riboflavin 5′-phosphate	Chemical treatment of vitamin B_2	Yellow colouring	Often used to fortify processed foods, especially breakfast cereals
E102	Tartrazine	One of a group of synthetic chemicals originally derived from coal tar, known as azo dyes	Yellow colouring	One of the commonest food colourings. Found in very wide range of foods, not just yellow ones. Breadcrumbs, tinned peas, broad beans, green beans, fish fingers, fish cakes, fish

crumbles, smoked fish, breaded fish, scallops, range of cakes, tarts, Swiss rolls, biscuits & sweets, cake mixes, cake decorations e.g. jelly diamonds, coloured sugar strands, mixed peel, marzipan, ice-cream, ice-cream wafers & cones, custard powder, custard type fillings, non-dairy cream, coffee whitener, packet soups, instant soups, packet savoury rice, pot casseroles, packet pudding mixes – various flavours, lemon & raspberry flavour mousses, tinned fruit-pie fillings, jellies, quick-set jells, gravy granules, instant sauces, cheese sauce mixes, fruity sauces, strawberry & banana milk shake syrups, tandoori paste, mint sauce, pickles, piccalilli, salad cream, crisps, cocktail cherries, instant hot chocolate drink, wide range of soft drinks including bitter lemon, ginger ale, orange & lemon squashes, blackcurrant drinks, etc., green & yellow food colouring, brandy flavouring

Code	Names	Source	Uses	Examples
E104	Quinoline yellow	Coal tar dye	Greenish-yellow colouring	Lemon curd, lemon fillings, strawberry flavour milk shake syrups, crisps, soft drinks & squashes

UK	EEC		WHO/FAO			Banned Abroad			Comments	Risk of/to	Rating	Code
Only Provisionally Approved/Tests Called For	Approved (No. of Food Categories Permitted in)	Being Considered	ADI Established	Only Temporary ADI/Tests Called For	No ADI	Nowhere	In 1-3 Countries	In 4 or More Countries				
○	○			○		○			Curcumin can damage animal genes. FACC has expressed doubts about it	Conception, ?cancer	△ △ △ △	**E100**
	○		○			○			An important nutrient, naturally present in many foods; can be metabolised to vitamin B_2; often used to fortify processed foods stripped of their nutrients		√√	**E101**
		○	○					○			√√	**101a**
	○		○			○			One of the most widely used of dyes. Despite good toxicity data, it is known to provoke reactions in lungs, nose, skin & eyes, & to trigger hyperactivity. Those who are sensitive to salicylates/aspirin are especially likely to suffer adverse reactions. Adverse effects occur at doses well below official ADIs. Safety margins poor	Hyperactivity, asthma, eczema, hives, anyone sensitive to salicylates	△ △	**E102**
○	○		○			○			Like E102, can provoke allergic reactions. FACC called for further safety tests. Safety margins borderline	As E102; hypersensitivity, hyperactivity	△ △	**E104**

Code	Names	Source	Uses	Examples
107	Yellow 2G	Azo dye (see E102)	Yellow colouring	Tinned vegetables, sweets, chocolate, pickles, sauces
E110	Sunset yellow FCF; Orange yellow S	Azo dye (see E102)	Orange-yellow colouring	Yoghurts, jams, lemon curd, tinned butter beans, fish cakes, fish fingers, smoked haddock, salami, range of cakes, tarts & Swiss rolls, cake decorations e.g. jelly diamonds, coloured sugar strands, marzipan, variety of biscuits, wide range of sweets, ice-cream, custard powder, custard-type filling, non-dairy cream, coffee whitener, packet savoury rice, pot casseroles, packet pudding mixes – various flavours, lemon meringue fillings, jellies, cheese sauce mixes, fruity sauces, milk shake syrups tomato ketchup, tandoori paste, lobster & crab spread crisps, savoury onion rings, cocktail cherries, instant hot chocolate drink, soft drinks & squashes, brandy flavouring
E120	Cochineal; Carminic acid	Cactus beetles	Red colouring	Rare because expensive. Garlic sausage, alcoholic drinks
E122	Carmoisine, Azorubine	Azo dye (see E102)	Red colouring	Wide range of foods: yoghurts, jams, bramble jelly, pickled beetroot, range of cakes & iced fancies, cake decorations e.g. jelly diamonds, coloured sugar strands, marzipan, wide range of sweets including chocolate drops, marshmallows, ice-creams, packet soups, curried packet rice, packet pudding mixes – various flavours, tinned fruit pie fillings, raspberry & chocolate flavoured mousses, jellies, red cherry topping on cheesecakes, instant sauces, gravy granules, fruity sauces, syrups, so-called blackcurrant 'health' drinks, brandy flavouring

UK			EEC		WHO/FAO			Banned Abroad			Comments	Risk of/to	Rating	Code
Approved	Only Provisionally Approved	Approved/Tests Called For	Approved (No. of Food Categories Permitted in)	Being Considered	ADI Established	Only Temporary ADI/Tests Called For	No ADI	Nowhere	In 1–3 Countries	In 4 or More Countries				
	○			○						○	Suspected carcinogen. Also provokes hypersensitive reactions. Affected the kidneys of animals in tests (FACC)	?Cancer & as E102; hypersensitivity	△ △ △	**107**
○			○(15)		○				○		Suspected carcinogen. Can provoke hypersensitive reactions; & can make people allergic to sunlight (FACC)	?Cancer & as E102; hypersensitivity	△ △ △	**E110**
	○		○(5)			○			○		Toxic to the embryo. Provokes hypersensitive reaction in children. FACC expressed doubts about its safety	Conception, hyperactivity, hypersensitivity	△ △	**E120**
	○		○(11)		○				○		Suspected carcinogen & mutagen. Provokes hypersensitive reactions. FACC expressed doubts about its safety. Safety margins poor	Conception, ?cancer, hypersensitivity	△ △ △	**E122**

Colourings E123–E127

Code	Names	Source	Uses	Examples
E123	Amaranth	Azo dye (see E102)	Purple-red colouring	One of the most widely used colourings in the UK for: breadcrumbs, black cherry yoghurt, pickled beetroot, jams, chocolate cakes, Swiss rolls, tarts, pot rice, instant soups, raspberry flavour mousses, jellies, quick-set jells, tinned fruit-pie fillings, ice-cream, ice-pops, tomato ketchup, strawberry flavour syrups, soft drinks including cherryade, blackcurrant drinks, squashes, as well as vitamin C & other so-called 'health' drinks, red food colouring, brandy flavouring
E124	Ponceau 4R, Cochineal red A	Azo dye (see E102)	Red colouring	Strawberry yoghurt, strawberry jam, tinned raspberries & strawberries, fish cakes, breaded fish, salami, beefburgers, variety of cakes & gateaux, doughnuts & chocolate Swiss rolls, cake mixes, mixed peel, cocktail cherries, glacé cherries, wide range of biscuits, ice-cream, ice-cream wafers & cones, ice-pops, custard type fillings, instant soups, packet soups, frozen pizza, tinned fruit-pie filling, lemon meringue pie filling, red quick-set jells, dessert toppings, tomato ketchup, cheese sauce mix, fruity sauces, mango chutney, tandoori paste, soft drinks & squashes, red & blue food colouring
E127	Erythrosine BS	Coal tar dye	Red colouring	Tinned fruit cocktail, rhubarb & strawberries, salami, sausages, sausage meat, luncheon meat, cherry cakes, gateaux, Swiss rolls, cake decorations e.g. jelly diamonds, coloured sugar strands, glacé cherries, iced biscuits, chocolate covered wafers & biscuits, wide range of sweets including Turkish delights & marshmallows, custard powder, frozen pizza, tinned fruit-pie fillings, lobster, crab & beef spreads, savoury onion rings

UK	EEC		WHO/FAO			Banned Abroad			Comments	Risk of/to	Rating	Code
Only Provisionally Approved/Tests Called For	Approved (No. of Food Categories Permitted in)	Being Considered	ADI Established	Only Temporary ADI/Tests Called For	No ADI	Nowhere	In 1–3 Countries	In 4 or More Countries				
○	○(11)			○				○	Suspected carcinogen, may damage the foetus. Mutagen. Inadequate safety data. FACC has called for more. Provokes skin allergies. Safety margins poor	Conception, ?cancer, hypersensitivity	△△△	**E123**
○	○(12)	○						○	3 colours closely related to E124 were banned worldwide in 1965. Suspected carcinogen. Like E102 provokes hypersensitive reactions. FACC called for more tests	Conception, ?cancer, hypersensitivity	△△△	**E124**
○	○(11)	○						○	Suspected carcinogen. Can provoke hypersensitivity & allergic reaction to light. Suspected mutagen. FACC unhappy with safety data. Safety margins poor	Conception, ?cancer, hypersensitivity, thyroid disease	△△△	**E127**

Colourings 128–E141

Code	Names	Source	Uses	Examples
128	Red 2G	Azo dye (see E102)	Red colouring	Coating on fish fingers, sausage garlic & smoked sausages, frankfurters, Scotch eggs, chocolate rolls, chocolate flav dessert whips, soft drinks
E131	Patent blue V	Coal tar dye	Blue colouring	Peas, fruit flavoured sweets, ice-cream, Scotch eggs
E132	Indigo carmine, Indigotine	Coal tar dye	Dark-blue colouring	Cake decorations e.g. jelly diamonds, coloured sugar strands, chocolate flavoured coating, iced biscuits, chocola covered wafers, range of swe instant sauces, savoury onion rings, brandy flavouring. Also used in hospitals to check kid function
133	Brilliant blue FCF	Coal tar dye	Blue colouring often combined with E102 to give green	Tinned peas, sweets, biscuits ice-cream, jellies – grapefruit pineapple flavour, sparkling fr drinks, apple & blackcurrant d blue food colouring
E140	Chlorophyll	Natural leaf pigment extracted from nettles, grass or alfalfa	Green colouring	Fats, preserved vegetables
E141	Chlorophyllins	Chemical treatment of E140	Green colouring	Cucumber relish, preserved vegetables

UK Only Provisionally Approved/Tests Called For	EEC Approved (No. of Food Categories Permitted in)	EEC Being Considered	WHO/FAO ADI Established	WHO/FAO Only Temporary ADI/Tests Called For	WHO/FAO No ADI	Banned Abroad Nowhere	Banned Abroad In 1-3 Countries	Banned Abroad In 4 or More Countries	Comments	Risk of/to	Rating	Code
○		○	○					○	Suspected carcinogen & mutagen. Banned in several countries. FAO have called for extensive study into risks of birth defects, & effects on bone & red blood cell formation. FACC have expressed concern about safety. Breaks down to Red 10B which is thought to cause chromosomal damage	Conception, ?cancer, hypersensitivity	△ △ △	**128**
○	○(2)			○				○	Suspected carcinogen. 'Available evidence on all aspects is inadequate,' (FACC 29th report). Nausea, low blood pressure, shaking reported. Provokes hypersensitivity	?Cancer, hypersensitivity, nervous disorders	△ △ △	**E131**
	○(10)	○					○		Suspected carcinogen & acknowledged as such by US FDA. Sickness, raised blood pressure, hypersensitive reactions reported	?Cancer, hypersensitivity	△ △ △	**E132**
			○	○				○	Suspected carcinogen. Can provoke allergic reactions	?Cancer, hypersensitivity	△ △ △	**133**
	○		○			○			Seems safe		√√	**E140**
	○		○			○			Seem safe		√√	**E141**

Code	Names	Source	Uses	Examples
E142	Green S, Acid brilliant green, Lissamine green	Coal tar dye	Green colouring	Variety of jams, e.g. plum, blackcurrant, bramble, tinned soups, peas, broad beans, green beans, range of cakes, gateaux, chocolate covered Swiss rolls, mixed peel, fruit flavoured sweets, soft-centred sweets, chocolate ice-cream, ice-pops, tinned fruit-pie fillings, chocolate flavoured mousses, jellies, gravy granules, mint sauce, soft drinks including bitter lemon, lemon lime drinks & fruit cocktails, as well as blackcurrant 'health' drinks, green food colouring
E150	Caramels	Chemically 100 ways, none of products definable. Most forms are made with ammonia. Number in use being reduced to 6	Brown colour & flavour	98% by weight of all added colouring in the UK, covering a enormous range of products: flour, bread, marmalade, tinned barbecue beans, breaded fish, fish cakes, very wide range of tinned and prepared meat products, hamburgers, meatballs, steak & kidney pie, meat pasties, tinned meat pie fillings, mince meat, Scotch eggs, sponge cakes, meat loaf, gateaux, chocolate rolls, doughnuts, packet cake mixes, mincemeat, range of biscuits & chocolate covered wafers, chocolate, chocolate ice-cream, tinned oxtail soup, packet soups, range of sauces in tins, packet savoury rice dishes, pot rice, tinned curries, packet curries, packet pudding mixes, stock cubes, gravy granules, gravy mixes, soya sauce, fruit sauces, pickles, relishes, pickled gherkins, French dressing, peach chutney, beef paste, ginger ale, cola, beer, some whiskies, rum flavouring, vanilla essence
E151	Black PN, Brilliant black BN	Azo dye (see E102)	Dull purple colouring	Blackcurrant products, black cherry & damson jam, gravy granules, fruity sauces
E153	Carbon black, Vegetable carbon	Burnt plant material	Black colouring	Chocolate flavoured coating, choc ices, chocolate flavoured dessert mixes, fruit juice concentrates, jams, sweets

UK	EEC			WHO/FAO		Banned Abroad						
Only Provisionally Approved/Tests Called For	Approved (No. of Food Categories Permitted in)	Being Considered	ADI Established	Only Temporary ADI/Tests Called For	No ADI	Nowhere	In 1–3 Countries	In 4 or More Countries	Comments	Risk of/to	Rating	Code
○	○(11)				○			○	Banned in UK before entry to EEC. Allergic reactions reported. Safety margins poor	Hypersensitivity	△△	**E142**
○	○			○			○		Long in doubt, inadequately researched. Some forms are mutagenic and are banned in some countries. Officials admit intake liable to exceed ADI. Considered a natural additive for the purposes of labelling, even though most sorts made with ammonia these days. Some types produce vitamin B_6 deficiency in animals. FACC has called for more safety tests. When used as a flavouring (as it often is) need not be declared separately	Conception, ?cancer	△△△	**E150**
○	○(4)	○						○	FACC admit testing is inadequate. Safety margins borderline. Caused intestinal cysts when fed to pigs. Provokes hypersensitive reactions	Hypersensitivity, bowel disorders	△△	**E151**
	○				○		○		Suspected carcinogen. Burnt material contains mutagenic compounds. Linked with skin cancer in workers handling large amounts	Conception, ?cancer	△△△	**E153**

Code	Names	Source	Uses	Examples
154	Brown FK, Kipper brown, Food brown	Azo dye mixture (see E102)	Yellow-brown colouring	Used in a wide range of brown foods: smoked mackerel, kippers, pot rice, sausages, preserved meats, gravy granules, crisps, sweets, biscuits
155	Brown HT, Chocolate brown HT	Azo dye (see E102)	Brown colouring	Chocolate products, chocolate covered cakes, Swiss rolls, gateaux, chocolate ice-cream
E160(a)	Alpha-carotene, Beta-carotene, Gamma-carotene	Natural plant pigment extracts	Yellow colouring	Used increasingly due to consumer pressure for safe colourings: margarine, low fat spread, cheeses – Edam, Double Gloucester, Red Cheshire, Red Leicester, Port Salut, cheese spread, cheese flan, dried scalloped potatoes, ploughman pasties, range of cakes & biscuits, sweets, peanut sweets non-dairy cream, tinned lentil & celery soup, tinned macaroni cheese, wine sauces in tins, frozen pizzas, cheese sauce granules, raspberry & chocolate flavour mousses, packet pudding mixes – all flavours, cheesecake mixes, frozen pastry, packet dessert toppings
E160(b)	Annatto; Bixin; Norbixin	Seed pods of annatto tree	Yellow colouring	
E160(c)	Capsanthin; Capsorubin	Paprika	Orange colouring	
E160(d)	Lycopene	Tomatoes	Red colouring	
E160(e)	Beta-apo-8'-carotenal, Beta-8'-apocarotenal	Plant extract	Orange colouring	
E160(f)	Ethyl ester of E160(e)	Plant derivative	Orange-yellow colouring	

| Only Provisionally Approved/Tests Called For | Approved (No. of Food Categories Permitted in) | Being Considered | ADI Established | Only Temporary ADI/Tests Called For | No ADI | Nowhere | In 1–3 Countries | In 4 or More Countries | Comments | Risk of/to | Rating | Code |
UK	EEC		WHO/FAO			Banned Abroad						
○		○			○			○	Banned in several countries because causes mutations. May cause cancer. The FK is 'for kippers', but appears in a very wide range of products, so safety margins disturbingly low. FACC expressed doubts about its safety	Conception, ?cancer, hypersensitivity	△ △ △	154
○		○		○				○	Suspected carcinogen, banned in US as such. FACC has expressed concern about it. WHO/FAO has called for more work on the risks of birth defects	Conception, ?cancer, hypersensitivity	△ △ △	155
		○			○	○					√√	E160(a)
○	○			○		○			Solvent extracts still subject to safety testing. Water extracts considered entirely safe. Annatto can cause allergic reactions. Lycopene may be banned by EEC. Some are effective as vitamin A		△	E160(b)
	○				○	○					√√	E160(c)
	○				○	○					△	E160(d)
	○		○					○			√√	E160(e)
	○		○			○					√√	E160(f)

Colourings E161–E163(f)

Code	Names	Source	Uses	Examples
E161	Xanthophylls	Carotenoids	Yellow colourings	
E161(a)	Flavoxanthin	Buttercups	Yellow colouring	
E161(b)	Lutein	Leaves, egg yolk	Yellow colouring	
E161(c)	Cryptoxanthin	Various plants, orange rind, egg yolk, butter	Yellow colouring	Found in quantity in eggs & chicken fat, undeclared
E161(d)	Rubixanthin	Rosehips	Yellow colouring	
E161(e)	Violaxanthin	Yellow pansies	Yellow colouring	
E161(f)	Rhodoxanthin	Yew tree seeds	Yellow colouring	
E161(g)	Canthaxanthin	Mushrooms, crustaceans, trout, salmon, tropical birds	Orange colour	Artificial sun tan capsules. 'Addition to foods not recommended' (FACC)
E162	Beetroot red, Betanin	Beetroot	Purple-red colouring	Not widely used
E163	Anthocyanins			
E163(a)	Cyanidin		Red colouring	
E163(b)	Delphinidin		Blue colouring	
E163(c)	Malvidin	Natural pigments from various plants	Purple colouring	Use limited because of expens
E163(d)	Pelargonidin		Brown colouring	
E163(e)	Peonidin		Red colouring	
E163(f)	Petunidin		Red colouring	Glacé cherries, blackcurrant dr.

UK		EEC		WHO/FAO			Banned Abroad			Comments	Risk of/to	Rating	Code
Approved	Only Provisionally Approved/Tests Called For	Approved (No. of Food Categories Permitted in)	Being Considered	ADI Established	Only Temporary ADI/Tests Called For	No ADI	Nowhere	In 1-3 Countries	In 4 or More Countries				
												√	E161
												√	E161(a)
												√	E161(b)
												√	E161(c)
○		○					○	○		Research inadequate. Found in quantity in eggs & chicken fat, because added to feed, where it is undeclared		√	E161(d)
												√	E161(e)
												√	E161(f)
○		○					○		○			△	E161(g)
○		○					○	○		Seems safe		√√√	E162
												√√	E163
												√√	E163(a)
												√√	E163(b)
○		○					○		○	Seem safe		√√	E163(c)
												√√	E163(d)
												√√	E163(e)
												√√	E163(f)

Colourings E170–E180

Code	Names	Source	Uses	Examples
E170	Calcium carbonate, Chalk	Occurs naturally as a mineral	Alkali firming agent, release agent, calcium supplement, surface colouring	White flour, white bread flour, therefore also in biscuits, bread, cakes, etc., instant porridge, sweets, ice-cream, vitamin pills & tablets
E171	Titanium dioxide	Mineral sources	White surface colouring	Used primarily in confectionery, pickles, sauces, cottage cheese - salmon & cucumber flavour, fruit flavour sweets, pot rice, cheese sauce granules, packet curried rice, horseradish, pills
E172	Iron oxides, Iron hydroxides	Mineral pigments	Colouring (various)	Packet curried rice, crab, salmon & shrimp pastes & spreads, cocoa paste
E173	Aluminium			
E174	Silver	Metals	Colour coating	Rarely used. Pills, sugar-coated confectionery
E175	Gold			
E180	Pigment rubine, Lithol rubine BK	Azo dye (see E102)	Red colouring	Cheese rind only

UK		EEC		WHO/FAO			Banned Abroad			Comments	Risk of/to	Rating	Code
Approved	Only Provisionally Approved/Tests Called For	Approved (No. of Food Categories Permitted in)	Being Considered	ADI Established	Only Temporary ADI/Tests Called For	No ADI	Nowhere	In 1-3 Countries	In 4 or More Countries				
○		○		○					○	Seems safe. However, often used to supplement food which would otherwise be of poor nutritional quality		√√	**E170**
○		○(3)		○					○	Inert, considered very safe by FACC, but little testing done		√	**E171**
○		○(1)		○					○	Considered very safe by FACC, but little toxicity testing		√√	**E172**
○		○(1)			○				○	Banned in several other countries, but little used. E173 can trigger hypersensitivity & is toxic. E174 can colour skin permanently & is toxic. Allergy to E175 is rare	Hypersensitivity	△△	**E173**
											Hypersensitivity	△△	**E174**
												√	**E175**
○		○(1)			○			○	○	Provokes hypersensitive reaction. Only on cheese rind, so can at least be removed, but there is a slight risk of contamination	Hypersensitivity	△△	**E180**

Preservatives (E200–297)

Preservatives are frequently cited as justification for all food additives: without additives we would keel over from food poisoning and vast quantities of food would be wasted, the argument goes. The fact is that preservatives make up a tiny proportion of all the additives used (less than 1 per cent); and there are many equally effective and considerably safer methods of keeping food (e.g. refrigeration, pasteurisation, better hygiene in factories). Moreover, several manufacturers manage quite happily without them (see our list of alternative products, p. 243). But preservatives have other advantages – they enable manufacturers to prolong shelf life long past the normal life of the food, they may have useful cosmetic properties (see nitrates), or they may be needed to preserve other additives being used.

The chemical characteristics which make preservatives effective in preserving food also make them potentially harmful to humans. Many of those listed in this section are cause for serious concern.

Because of the hazards associated with them, all those additives officially classed as preservatives are restricted in this country – the levels at which they may be used and the foods to which they may be added are both limited. WHO/FAO has also suggested restrictions. Nevertheless, they are used very widely and in large quantities by comparison with most colourings. It is easy to consume them in considerable doses from several different sources if you eat a lot of processed food – the safety limits do not take this into account fully. Old-fashioned methods of preserving tend to deal more or less honestly with our senses. They spoil the appearance, texture and flavour of the preserved food as they reduce its nutritive quality so that we are not much misled about them. Twentieth century developments have meant an enormous number of foods can now be kept bacteriologically safe very easily, and this generally means that they have very little nutritional value. Moreover, it has meant most people in this country today consume far less fresh food and as a result far fewer of the vitamins, minerals and essential fats they need.

Preservatives E200–E219

Code	Names	Source	Uses	Examples
E200	Sorbic acid	Mountain ash berries, other fruits. Can all be synthesised	To inhibit growth of yeasts, moulds & some bacteria	Very wide range of products: low fat spreads, cottage cheese, processed cheese slices, cheese spreads, yoghurts, sour milk,
E201	Sodium sorbate			
E202	Potassium sorbate	wide range of cakes, tarts, fruit pies, Swiss rolls, fruit for ice-cream, candied peel, glacé fruit, dried figs, jams & pickles, frozen pizzas, snack meals, cheesecake mixes, desserts, thousand island dressing, soya sauce, garlic & herb dressing, cucumber relish, fish patés. Also in many drinks, including cider, wine, apple juice & soft drinks & to preserve other additives, e.g. colourings & anti-foaming agents		
E203	Calcium sorbate			
E210	Benzoic acid	Cranberries & other edible berries. Usually synthesised	To inhibit bacterial growth	Wide variety of food & drinks, bakery products, pizzas, marinated herring, yoghurt, cheese, low fat spreads, blue
E211	Sodium benzoate	cheese dressing, fruit products, jams, figs, nut paste, pickles and sauces (e.g. horseradish), salad cream, sweets, candied peel, glacé fruit, gelatin, desserts – toppings & fillings, milk shake syrups, fruit ice-cream. Also beer, cider, wine, perry, ginger ale, ginger beer, shandy, tonic water, barley water, squashes, colas, glucose drinks, cherryade, sparkling canned drinks, soda stream concentrates, frozen fruit juices. Frequently used to preserve other additives, e.g. flavourings, colourings, anti-foaming agents, artificial sweeteners, & so not necessarily declared		
E212	Potassium benzoate			
E213	Calcium benzoate			
E214	Ethyl 4-hydroxy-benzoate, Ethyl parahydroxybenzoate	Synthetic	Antimicrobial	Very wide range of products: fruit juices, glacé & candied fruit, jams, yoghurts, soft drinks, glucose drinks, tinned soups & concentrates, snack meals, pickles, sauces, salad cream, marinated herring. Also popular as preservative for other additives especially colourings, flavourings, artificial sweeteners, anti-foaming agents, foam heads (e.g. in beer), & so not necessarily declared
E215	Sodium salt of E214			
E216	Propyl 4-hydroxy-benzoate, Propyl parahydroxybenzoate			
E217	Sodium salt of E216			
E218	Methyl 4-hydroxy-benzoate, Methyl parahydroxybenzoate			
E219	Sodium salt of E218			

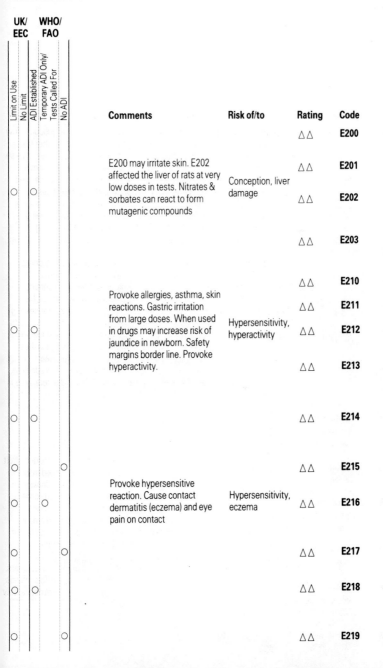

Limit on Use	No Limit	ADI Established	Temporary ADI Only/Tests Called For	No ADI	Comments	Risk of/to	Rating	Code
							△△	**E200**
					E200 may irritate skin. E202 affected the liver of rats at very low doses in tests. Nitrates & sorbates can react to form mutagenic compounds	Conception, liver damage	△△	**E201**
○		○					△△	**E202**
							△△	**E203**
							△△	**E210**
					Provoke allergies, asthma, skin reactions. Gastric irritation from large doses. When used in drugs may increase risk of jaundice in newborn. Safety margins border line. Provoke hyperactivity.	Hypersensitivity, hyperactivity	△△	**E211**
○		○					△△	**E212**
							△△	**E213**
○		○					△△	**E214**
○				○			△△	**E215**
○			○		Provoke hypersensitive reaction. Cause contact dermatitis (eczema) and eye pain on contact	Hypersensitivity, eczema	△△	**E216**
○				○			△△	**E217**
○		○					△△	**E218**
○				○			△△	**E219**

Preservatives E220–E232

Code	Names	Source	Uses	Examples
E220	Sulphur dioxide	Burning of sulphur or gypsum	Antioxidant, improver, bacteriocide, bleach (flour), vitamin C – preservative	Very wide range of foods, often large quantities (e.g. 2,000 ppm in dried fruit), also as additives to other additives, e.g. flavourings, colourings, anti-foaming agents, modified & hydrolised starches; cereals, such as muesli, jams, marmalades, pickles, packet scalloped potatoes, instant mash potato, dried vegetables, dried fruit, mincemeat, mixed peel, glacé cherries, cocktail cherries, beefburgers, sausages & sausage meat, turkeyburgers, tinned prawns, fresh (sic) seafood, packet & instant soups, packet savoury rice, pot casseroles, pot rice, packet stuffing, milk shake syrup, yoghurts, lemon juice, fruit juice, pickled onions, relishes, horseradish sauce, salad cream, vinegar, garlic powder, root ginger. Also in wide range of drinks, including squashes, blackcurrant drinks, barley water, bitter lemon, ginger ale, shandy, beer, cider, wine
E221	Sodium sulphite			
E222	Sodium hydrogen sulphite, Sodium bisulphite, Acid sodium sulphite			
E223	Sodium metabisulphite, Disodium pyrosulphite	Synthetic	Antioxidant, antimicrobial	
E224	Potassium metabisulphite, Potassium pyrosulphite			
E226	Calcium sulphite		Fermentation preventer	
E227	Calcium hydrogen sulphite, Calcium bisulphite			
E230	Biphenyl, Diphenyl			
E231	2-Hydroxy-biphenyl, O-phenyl phenol, Orthophenyl phenol	Synthesised from benzene	Antifungal	Skin of citrus fruits
E232	Sodium biphenyl-2-yl oxide, Sodium orthophenyl phenate, Sodium orthophenyl phenol			

Restrictions on Use

Limit on Use	No Limit	ADI Established	Temporary ADI Only/Tests Called For	No ADI	Comments	Risk of/to	Rating	Code
O		O			E220 has caused several deaths in USA; it is well		△ △ △	**E220**
	O	O			documented that it can be fatal to mild asthmatics. Extremely		△ △ △	**E221**
					common. Many reports of adverse reactions – weakness, shortness of breath, loss of consciousness. 115 ppm fed to rats killed half of them. Mutagenic – FAO called for extensive research (1978). All		△ △ △	**E222**
					sulphites act as sulphur dioxide ultimately. Destroys vitamin B_1. Intestinal irritant;	Conception, ?cancer, hypersensitivity	△ △ △	**E223**
					provokes hypersensitive reactions. Interacts with other drugs to increase incidence of tumours in animals. Easy to consume worrying amounts because used so widely &		△ △ △	**E224**
O				O	permitted in high concentrations, e.g. dried fruit		△ △ △	**E226**
	O	O			may contain 2,000 mg/kg. Can be removed from fruit by soaking & cooking, but consumers of raw dried fruit should choose unsulphured products		△ △ △	**E227**
O		O					△ △ △	**E230**
O		O			Used on peel of citrus fruit, but may contaminate fruit itself & products containing peel. E230/231 suspected carcinogens; E231 suspected mutagen. Can cause nausea, vomiting, eye & nose irritation, limb & abdominal pain, liver		△ △ △	**E231**
O				O	damage. 1 out of 9 workers died after accidental exposure; the other 8 were all seriously ill. Safety margins poor. Wash hands after peeling citrus fruits. Wash peel in dilute detergent, then rinse	Conception, ?cancer, hypersensitivity, rhinitis	△ △ △	**E232**

Code	Names	Source	Uses	Examples
E233	Thiabendazole, 2(thiazol-4-yl) benzimidazole	Synthetic	Fungicide	Banana & citrus skins, so can contaminate products using the peel, e.g. marmalade. We also found it actually declared in one orange squash
234	Nisin	Natural product of several cheese-starter bacteria	Antibiotic	Cheese spread & other cheese
E236	Formic acid	Natural from ants, or synthetic	Antibacterial	Not permitted in UK but may be imported indirectly
E237	Sodium formate			
E238	Calcium formate			
E239	Hexamine, Hexamethylene tetramine	Synthesised from 'benzene'	Fungicide	Provolone cheese, marinated herring/mackerel

	UK/EEC		WHO/FAO					
Limit on Use	No Limit	ADI Established	Temporary ADI Only/Tests Called For	No ADI	Comments	Risk of/to	Rating	Code
O				O	When used medicinally at 150 mg/kg of body weight produces wide range of adverse effects in many people. Safety margins poor	Intestinal, nervous & blood disorders, high blood pressure	△	**E233**
	O	O			No adverse effects reported. Being considered for acceptance by EEC		√	**234**
							△ △ △	**E236**
	O			O	Not permitted in UK in any human foods, but may be imported indirectly. Acid burns the skin, all can upset the kidneys		△ △ △	**E237**
							△ △ △	**E238**
O	O				Mutagenic & carcinogenic in animal experiments, even though occurs in natural rotting of tissues. Becomes formaldehyde in the gut; irritant. May affect kidneys, 'should not be mixed with nitrites,' (FAO/WHO)	?Cancer, kidney disease, conception	△ △ △	**E239**

Preservatives E249–E263

Code	Names	Source	Uses	Examples
E249	Potassium nitrite	Natural mineral	Curing agents, preservatives especially against botulism. Give attractive pink colour to meats containing them	Wide range of meat products: sausages, hot dogs, salami, garlic sausage, ham, tinned pork, corned beef, luncheon meat, bacon, gammon steaks, spam, meat pastes, paté, chicken roll. Edam cheese, cheese flans, frozen pizzas
E250	Sodium nitrite	Synthetic		
E251	Sodium nitrate, Chile saltpetre	Natural mineral		
E252	Potassium nitrate	Synthetic		
E260	Acetic acid	Acid distilled from wood, natural in vinegars. Others prepared chemically from acid	Antibacterial stabiliser, colouring diluent	
E261	Potassium acetate		colour preservative, buffer	Bread, cheese, frozen vegetables, fruit pastilles, packet soups, quick-set jells, crisps, salad cream, pickles
E262	Sodium hydrogen acetate, Sodium diacetate		Antibacterial, especially against spore-bearing bacteria.	
262	Sodium acetate		Antibacterial	
E263	Calcium acetate		Firming agent, mould suppressant, sequestrant	

	UK/EC			WHO/FAO				Comments	Risk of/to	Rating	Code
No Limit	ADI Established	Temporary ADI Only/Tests Called For	No ADI								
					○			Some of the most worrying of additives commonly used. FACC recommended in 1978 that their use be eliminated from food 'as soon as practicable'. Widely consumed at around 90mg per day. 4/5ths of that from food. But quantity varies greatly with diet.	Conception, ?cancer, blood disorders;	△ △ △	**E249**
					○				babies & young children	△ △ △	**E250**
		○								△ △ △	**E251**
		○								△ △ △	**E252**

Vegetables grown on nitrates or under glass may reach 1,000 ppm easily. Level of nitrates in water already exceeds ECC maximum in some parts of UK, & is likely to increase as nitrates from fertilisers penetrate the water table. Nitrates can be converted to nitrates which may combine with amines from food and drugs in the stomach to form nitrosamines, known to be very powerful carcinogens. Nitrosamines are also formed during cooking. It is agreed by nearly all that they are a theoretical cancer hazard, though there has been controversy about practical evidence in man. High nitrate intake has been related to high incidence of stomach cancer. These additives are also known to be toxic, they prevent the blood carrying oxygenm, infants particularly at risk. Their function could be achieved more safely by alternative means in many products, but the additional cosmetic effects of nitrates are a great asset to the meat processors – they produce a pinkish colour – especially in raw meat products where colourings are not permitted. They are banned in foods for babies.

	UK/EC			WHO/FAO				Comments	Risk of/to	Rating	Code
○				○						√ √	**E260**
○				○				Limited only by good manufacturing practice. Acid kills bacteria at above 5% concentration in traditional home preserves. Deters their growth in lesser concentrations		√ √	**E261**
○				○						√ √	**E262**
○				○						√	**262**
○				○						√	**E263**

Preservatives E270–297

Code	Names	Source	Uses	Examples
E270	Lactic acid	From sour milk, sauerkraut. Manufactured by fermentation	Flavour, antioxidant enhancer, preservative, acid	Bread, cheeses, milk powder, infant milks, margarine, olives oil, frozen vegetables, pickled cabbage, salad dressing, sweet apple/pear nectar, soft drinks
E280	Propionic acid			
E281	Sodium propionate	Acid, a natural digestive product in herbivore animals. Salts prepared chemically	Antifungal. E282 used in bread as a supplement	Wide range of breads; white, brown, wholemeal, granary, baps, malt loaf, burger buns, range of cakes, buns, flan cases apple roll, Christmas pudding, frozen pizzas, dairy products
E282	Calcium propionate			
E283	Potassium propionate			
E290	Carbon dioxide	Product of yeast fermentation	Packaging gas, carbonation, inhibits aerobic organisms	Fizzy drinks from water to beer
296	Dl-malic acid, L-malic acid	Naturally from apples, pears; & synthesised	Acidulant, flavour	Frozen vegetables, instant soup savoury rice, sauces, peach/pear nectar & prunes, low calorie orange drink, pineapple juice
297	Fumaric acid	Product of fermentation in many natural circumstances	Acidulant, flavour, raising agent	Cheesecake mixes & other gelatin desserts, soft drinks, sweets

Restrictions on Use

UK/EEC		WHO/FAO						
Limit on Use	No Limit	ADI Established	Temporary ADI Only/Tests Called For	No ADI	Comments	Risk of/to	Rating	Code
	○	○			'Should not be used in foods for infants under 3 months old, as they have difficulty metabolising it' (FAO/WHO)	Babies & young children	△	**E270**
							√	**E280**
					Use limited only by good manufacturing practice, but E281 is reported to cause some kinds of migraine, E282 skin irritations in bakery workers. The baker's union has banned its use in the UK in pure form		△	**E281**
○		○				Migraine, eczema	△	**E282**
							√	**E283**
	○			○	Hastens the absorption of fluids & alcohol from the stomach, enhancing their effect. Increases stomach acid secretion	Peptic ulcers	√	**E290**
○		○			Being considered for approval by EEC. Quantity restrictions on some foods in UK. A possible irritant. It is not known whether infants can metabolise Dl-malic acid properly	Eczema; babies & young children	△	**296**
○		○			Being considered for approval by EEC. A possible irritant	Eczema	△	**297**

Antioxidants (E300–E321)

When they come into contact with oxygen in the air, oils and fats are 'oxidised' and this makes them go rancid. Antioxidants prevent this happening. Unprocessed oils contain a natural antioxidant in the form of vitamin E, but this is destroyed in processing, and so manufacturers replace it with synthetic antioxidants. BHA and BHT (E320 and E321) are the commonest in this group, and both are highly suspect. Next are the gallates, which are also suspect. Antioxidants are often used not just to prevent oils from going rancid but to stop other additives, such as artificial colourings or flavourings, decomposing – for example, in crisps. Because antioxidants give a longer shelf life, they also give bacteria more opportunity to grow, and so they may be used in conjunction with preservatives.

The use of antioxidants is restricted by regulations in recognition of the hazards they pose. All antioxidant activity is potentially valuable, because it sustains us powerfully against the many chemical stresses which age us. But some of the agents used are harmful in themselves, and harmless (but more expensive) alternatives are available.

Antioxidants E300–E309

Code	Names	Source	Uses	Examples
E300	L-ascorbic acid, Vitamin C	Natural & synthetic	Vitamin, antioxidant, browning inhibitor, improver in flour, colour preservative in meat	Range of breads, baps, rolls, instant porridge, frozen potato shapes, frozen vegetables, jams, sausages, French bread pizzas, soft drinks: tonic water, bitter lemon, fruit juices, milk solids, baby foods
E301	Sodium l-ascorbate	Synthetic	Vitamin, antioxidant, colour, preservative	Sausages, sausage meats, frankfurters, hot dogs, luncheon meat, ham, bacon, gammon steak, chicken roll, paté, ox tongue, chopped & shaped meat, frozen pizzas, frozen vegetables, stock cubes
E302	Calcium l-ascorbate			
E304	6-O-Palmitoyl-l-ascorbic acid, Ascorbyl palmitate	Synthetic		
E306	Tocopherols, Vitamin E	Natural extracts of various oils		
E307	Alpha-tocopherol, Vitamin E	Synthetic	Vitamin, antioxidant	Sausages, oils, frozen vegetables, frozen pizzas, dessert toppings
E308	Gamma-tocopherol			
E309	Delta-tocopherol			

Limit on Use	No Limit	ADI Established	Temporary ADI Only/ Tests Called For	No ADI	Comments	Risk of/to	Rating	Code
			UK/ EEC				WHO/ FAO	
	O	O			Vitamin C, recommended daily intake 20–60mg in UK. Large supplements are well tolerated, though over 600mg daily is diuretic. Some reports of kidney stones on high doses, but safety margins good. More than 1g daily has appreciable contraceptive effect. Prevents nitrosamine formation in the mouth & stomach		✓✓✓	**E300**
	O	O					✓✓✓	**E301**
	O			O	Wide limits of safety		✓✓✓	**E302**
	O	O			Restrictions on uses & quantities allowed in UK		✓✓	**E304**
							✓✓✓	**E306**
							✓✓✓	**E307**
	O	O			Wide limits of safety		✓✓✓	**E308**
							✓✓✓	**E309**

Antioxidants E310–E321

Code	Names	Source	Uses	Examples
E310	Propyl gallate, Propyl 3,4,5 trihydroxy-benzene			Wide range of products containing fat: oils & fats, breakfast cereals, chocolate flavour dessert topping, popcorn chewing gum, mayonnaise, orange squash, instant mashed potato. Also used but not declared in dairy products used to make other foods, including butter, dried cream & cheese, & for other additives such as colourings, flavourings, emulsifiers, stabilisers
E311	Octyl gallate	Synthetic	Antioxidant	
E312	Dodecyl gallate, Dodecyl 3,4,5 trihydroxy-benzene			
E320	Butylated hydroxyanisole, BHA			Two of the most commonly used additives in UK. They appear in wide range of products containing fat: oils & fats (e.g. lard, dripping, margarine), lemon curd, instant mashed potato, packet scalloped potatoes, cake, pastry & crumble mixes, mincemeat, sweet biscuits, cheese biscuits, sweets, chewing gum, popcorn, tinned, instant & packet soups, savoury rice, pot casserole, dried risotto, gravy granules, cheese sauce granules, gravy powder, stock cubes, chocolate flavour dessert toppings, cheesecake mixes, mousses, packet puddings – various flavours, horseradish sauce, crisps, soft drinks. Also used in vitamin A preparations – including those for use in baby food. Also used in butter intended for manufacture, & for other additives including colourings, flavourings, emulsifiers, stabilisers
E321	Butylated hydroxytoluene BHT	Synthetic	Antioxidant	

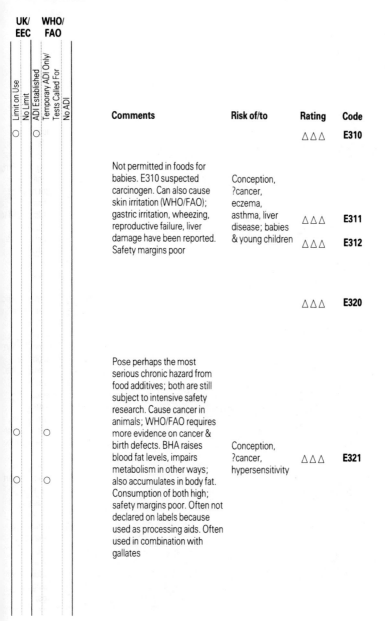

UK/EEC		WHO/FAO				Comments	Risk of/to	Rating	Code
Limit on Use	No Limit	ADI Established	Temporary ADI Only	Tests Called For	No ADI				
○			○					△ △ △	**E310**
						Not permitted in foods for babies. E310 suspected carcinogen. Can also cause skin irritation (WHO/FAO); gastric irritation, wheezing, reproductive failure, liver damage have been reported. Safety margins poor	Conception, ?cancer, eczema, asthma, liver disease; babies & young children	△ △ △	**E311**
								△ △ △	**E312**
								△ △ △	**E320**
○			○			Pose perhaps the most serious chronic hazard from food additives; both are still subject to intensive safety research. Cause cancer in animals; WHO/FAO requires more evidence on cancer & birth defects. BHA raises blood fat levels, impairs metabolism in other ways; also accumulates in body fat. Consumption of both high; safety margins poor. Often not declared on labels because used as processing aids. Often used in combination with gallates	Conception, ?cancer, hypersensitivity	△ △ △	**E321**
○			○						

Emulsifiers, Stabilisers, Thickeners and Others (E322–495)

Emulsifiers are used to bind together fat and water, which normally repel each other. Stabilisers stop them separating out again. Thickening agents sometimes act as emulsifiers and are classed in a group with them.

When you make mayonnaise, the natural lecithin in the egg acts as an emulsifier, binding the egg and oil together to make a thick dressing. The example is beloved of the food industry. Unfortunately, not all the commercially synthesised emulsifiers and stabilisers are so innocent. There are serious doubts about the safety of some of the additives in this group. Moreover, they are often used to disguise the lack of real ingredients in food. Ask yourself what they are doing in a product before you buy it. Are they making saturated fats and processed sugars, which are damaging to health, more attractive? Are they used to add worthless bulk to foods? The polyphosphates, for example, are frequently added to meats such as ham and sausages to make them soak up water, so that their weight increases, and they can be sold for more. The use of bulking aids is very common in slimming foods, where they are used to add body without calories.

Emulsifiers and stabilisers are regulated and there are some restrictions on foods to which they may be added, but there are few limits on the levels at which they may be used.

Emulsifiers, Stabilisers, Thickeners and Others E322–E331(c)

Code	Names	Source	Uses	Examples
E322	Lecithin	Soya beans, legumes, egg yolk	Emulsifier, stabiliser, antioxidant synergist, plasticiser	Bread, margarine, cakes, chocolate cake covering, chocolate drops, range of sweet & savoury biscuits, chocolate covered wafers, chocolate, popcorn, frozen pastry, ice-cream, ice-cream cones & wafers, cheese sauce granules, cheese spread, soft cheese, mayonnaise, cheesecake mixes, packet pudding mixes, dessert topping, milk shake mixes
E325	Sodium lactate	} Salts of lactic acid	Antioxidant synergist enhancer, buffer for lactic acid	Low fat spreads, meringue mixes, sweets. Also added to other additives, e.g. lactic acid, antioxidants
E326	Potassium lactate			
E327	Calcium lactate			
E330	Citric acid	Citrus juices, fermentation of molasses	Acid, antioxidant synergist, to prevent browning, to preserve vitamin C	A wide range of foods including: yoghurts, margarine, cottage cheese, jams, marmalades, lemon curd, tinned shrimps, prawns, crab, turkey burgers, cakes, fruit pies, doughnuts, sweets, ice-cream, instant soups, tinned sauces, packet sauce mixes, frozen pizzas, chocolate flavoured desert toppings, jellies, fruit-pie fillings, milk shake mixes, soft drink concentrates, crisps, soft drinks, sparkling drinks, wine, fruit juices, tinned fruit & vegetables, frozen vegetables, chips
E331	Sodium citrates	} Sodium salts of citric acid	Antioxidant synergist, buffer for citric acid, emulsifier	Low fat spreads, processed cheese slices, jams & marmalades, cakes, doughnuts, biscuits, sweets, savoury rice, jellies, mousses, drinks: soda water, diet drinks, cola, bitter lemon, wine
E331(a)	Sodium dihydrogen citrate			
E331(b)	Disodium citrate			
E331(c)	Trisodium citrate			

Restrictions on Use

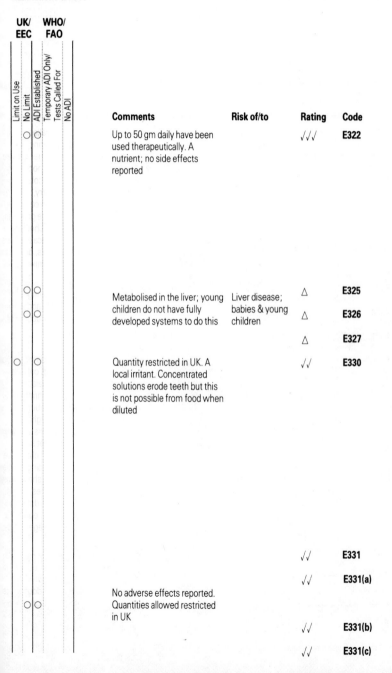

UK/EEC		WHO/FAO			Comments	Risk of/to	Rating	Code
Limit on Use	No Limit	ADI Established	Temporary ADI Only/Tests Called For	No ADI				
	O	O			Up to 50 gm daily have been used therapeutically. A nutrient; no side effects reported		√√√	**E322**
	O	O			Metabolised in the liver; young children do not have fully developed systems to do this	Liver disease; babies & young children	△	**E325**
	O	O					△	**E326**
							△	**E327**
O		O			Quantity restricted in UK. A local irritant. Concentrated solutions erode teeth but this is not possible from food when diluted		√√	**E330**
							√√	**E331**
							√√	**E331(a)**
	O	O			No adverse effects reported. Quantities allowed restricted in UK		√√	**E331(b)**
							√√	**E331(c)**

Emulsifiers, Stabilisers, Thickeners and Others E332–E336

Code	Names	Source	Uses	Examples
E332	Potassium dihydrogen citrate; Monopotassium citrate; Tripotassium citrate; Potassium citrate	Potassium salts of citric acid	Buffer for citric acid, emulsifier	Processed cheese, cheese spread, sweets, biscuits
E333	Calcium citrate; Monocalcium citrate; Dicalcium citrate; Tricalcium citrate	Calcium salts of citric acid	Buffer for citric acid, firming agent, emulsifier	Processed cheese, cheese spread, sweets, soft drinks
E334 —	L-(+)-tartaric acid; DL-tartaric acid	Grapes, wine	Antioxidant enhancer, aeration enhancer, flavour	Wine, sweets, jams, jellies, soft drinks
E335	Monosodium L-(+)-tartrate; Disodium L-(+)-tartrate; Monosodium DL-tartrate; Disodium DL-tartrate	Salts of tartaric acid	Sequestrant, antioxidant enhancer, emulsifier	Pizzas, sweets, jams, jellies, soft drinks
E336 — —	Monopotassium L-(+)-tartrate, Potassium hydrogen L-(+)-tartrate, Potassium acid L-(+)-tartrate, Cream of tartar; Dipotassium tartrate; Monopotassium DL-tartrate; Dipotassium DL-tartrate	Salts of tartaric acid	Sequestrant, antioxidant enhancer, emulsifier, raising agent	Wine, jelly crystals, meringue mixes

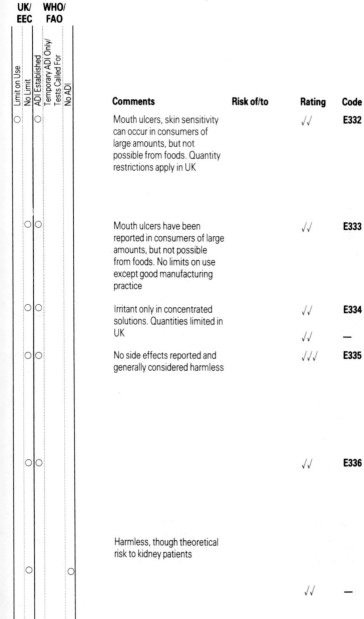

UK/EEC			WHO/FAO		Comments	Risk of/to	Rating	Code
Limit on Use	No Limit	ADI Established	Temporary ADI Only/Tests Called For	No ADI				
○		○			Mouth ulcers, skin sensitivity can occur in consumers of large amounts, but not possible from foods. Quantity restrictions apply in UK		√√	**E332**
	○	○			Mouth ulcers have been reported in consumers of large amounts, but not possible from foods. No limits on use except good manufacturing practice		√√	**E333**
	○	○			Irritant only in concentrated solutions. Quantities limited in UK		√√	**E334**
							√√	—
	○	○			No side effects reported and generally considered harmless		√√√	**E335**
	○	○					√√	**E336**
					Harmless, though theoretical risk to kidney patients			
	○			○				
							√√	—
							√√	—

Code	Names	Source	Uses	Examples
E337	Potassium sodium L-(+)-tartrate, Sodium potassium L-(+)-tartrate, Rochelle salt;	Salts of tartaric acid	Antioxidant enhancer, emulsifier, buffer for tartaric acid	Processed cheese, cheese spread, processed meats
—	Potassium disodium DL-tartrate			
E338	Orthophos-phoric acid, Phosphoric acid	Chemical treatment of phosphate rocks	Acid, flavour, antioxidant enhancer, setting aid, sequestrant	Sausages
E339(a)	Sodium dihydrogen orthophosphate Sodium hydrogen orthophosphate	Treatment of phosphoric acid	Raising agent, setting aid, buffer for phosphoric acid, stabiliser, emulsifier	Ham & sausages, low fat spreads, processed cheese slices, cheese spreads, sterilised cream, prepared fish dishes in cheese sauce, ice-cream, custard type fillings, tinned rice pudding, packet pudding mixes, cheesecake mixes, evaporated/UHT milk
E339(b)	Disodium hydrogen phosphate, Acid sodium phosphate			
E339(c)	Trisodium orthophosphate			

Restrictions on Use

UK/EEC		WHO/FAO						
Limit on Use	No Limit	ADI Established	Temporary ADI Only/Tests Called For	No ADI	Comments	Risk of/to	Rating	Code
	O	O					√√√	**E337**
					Little absorbed & harmless			
	O			O			√√√	—
O		O			Harmless in the quantities in food, but irritant in concentration. Quantities permitted are restricted in UK. General warning: phosphorus works with calcium and magnesium to produce healthy bones; too much phosphorus creates an imbalance; we generally eat too much phosphorus in the form of phosphates in processed foods		√√	**E338**
							√√	**E339(a)**
O		O			Harmless, & can be a nutrient, but see also general warning for E338		√√	**E339(b)**
							√√	**E339(c)**

Emulsifiers, Stabilisers, Thickeners and Others
E340(a)–E341(c)

Code	Names	Source	Uses	Examples
E340(a)	Potassium dihydrogen orthophosphate, Potassium phosphate monobasic			
E340(b)	Dipotassium hydrogen orthophosphate, Potassium phosphate dibasic	Treatment of phosphoric acid	Emulsifier, buffer for phosphoric acid, sequestrant, antioxidant enhancer	Custard type fillings, jellies, instant soups, dessert toppings, coffee whitener
E340(c)	Tripotassium orthophosphate, Potassium phosphate tribasic			
E341(a)	Calcium tetrahydrogen diorthophosphate, Acid calcium phosphate, ACP			
E341(b)	Calcium hydrogen orthophosphate, Calcium phosphate dibasic	Chemical treatment of natural mineral apatite	Raising agent, emulsifying agent, buffer, antioxidant enhancer, sequestrant, abrasive yeast nutrient	Bread, self-raising flour, baking powder, baked apple roll, cake mixes, mousse mixes
E341(c)	Tricalcium diorthophosphate, Tricalcium phosphate, TCP			
—	Ammonium dihydrogen orthophosphate, Ammonium phosphate monobasic			
—	Diammonium hydrogen orthophosphate, Ammonium phosphate dibasic	From phosphoric acid	Buffer, antioxidant enhancer, emulsifying salt	

UK/EEC		WHO/FAO						
Limit on Use	No Limit	ADI Established	Temporary ADI Only/Tests Called For	No ADI	Comments	Risk of/to	Rating	Code
							√√	E340(a)
○		○			Harmless & quantity limited in UK; but see warning for E338		√√	E340(b)
							√√	E340(c)
							√√	E341(a)
○		○			Harmless in quantities in food, which are restricted in UK. But see warning for E338		√√	E341(b)
							√√	E341(c)
	○			○			△	—
					Mildly diuretic, easily absorbed. When metabolised would release ammonia which is toxic to the nerves. Limits should be set	Nervous & kidney disorders	△	—

Code	Names	Source	Uses	Examples
350	Sodium malate; Sodium hydrogen malate			
351	Potassium malate	Treatment of malic acid, obtained from apples	Buffer for malic acid (296)	Products containing malic acid
352	Calcium malate; Calcium hydrogen malate			
353	Metatartaric acid	Treatment of tartaric acid	Sequestrant	Wine
355	Adipic acid, Hexanedioic acid	Widespread in nature, synthesised commercially	Raising agent, acidulant, flavour	Baking powder
363	Succinic acid	Treatment of acetic acid, occurs naturally in fungi	Acid	
370	1,4-Heptono-lactone	Synthetic	Acid, sequestrant	
375	Nicotinic acid, Niacin, Nicotinamide	Vitamin B_3, widespread naturally	Vitamin, colour, preservative	Bread, flour, breakfast cereals
380	Triammonium citrate	Treatment of citric acid	Buffer, emulsifier	

UK/EEC		WHO/FAO			Comments	Risk of/to	Rating	Code
Limit on Use	No Limit	ADI Established	Temporary ADI Only/Tests Called For	No ADI				
	○			○			√√	**350**
					Being considered for EEC approval. Seem safe		√√	**351**
	○	○					√√	**352**
○				○	Would be irritant in large quantities, so subject to restriction in UK. EEC is considering whether to approve it		√√	**353**
	○	○			Under EEC consideration. Probably irritant in quantity, but no adverse effects yet reported		√√	**355**
	○			○	Irritant in quantity, & no limits set. Under consideration for EEC approval		√	**363**
	○			○	Being considered for EEC approval. Irritant in quantity, & no limits set. Should be more carefully assessed		△	**370**
	○			○	Therapeutic quantities cause flushing, but unlikely in food use. Inclusion often required merely to replace losses in processing, & does not necessarily replace original nutritional value of raw food. Being considered by EEC		√√	**375**
	○	○			Could react with other food ingredients to release ammonia, which is irritant & toxic to nerves	Nervous disorders	△	**380**

Code	Names	Source	Uses	Examples
381	Ammonium ferric citrate, Ferric ammonium citrate, Ammonium ferric citrate, green	Treatment of citric acid	Iron supplement	Flour, pills, baby milk
385	Calcium disodium ethylene diamine-NNN'N'-tetra-acetate, Calcium disodium EDTA	Synthetic	Chelating agent	Tinned fish & shellfish, soft drinks, alcoholic drinks, glacé cherries
—	Disodium dihydrogen ethylene diamine-NNN'N' tetra-acetate, Disodium	Synthetic	Chelating agent	Brandy
E400	Alginic acid	Extract of seaweed harvested off Atlantic coast of British Isles	Stabiliser, gelling agent, emulsifier, thickener, foaming agent	Fish patés, cakes, doughnuts, sweets, ice-cream, custard type fillings, mousse mixes, cake mixes, stuffed olives, pimento paste, coleslaw, savoury onion rings, soft/processed cheese & spreads, soft drinks, beer
E401	Sodium alginate			
E402	Potassium alginate	Treatment of alginic acid	Emulsifier, stabiliser, gelling agent, thickener, solvent	Cakes, cottage cheese, cheese spread, blue cheese dressing, salad cream, mint sauce, squashes, beer. Also used as carrier for flavourings
E403	Ammonium alginate			
E404	Calcium alginate			

Restrictions on Use

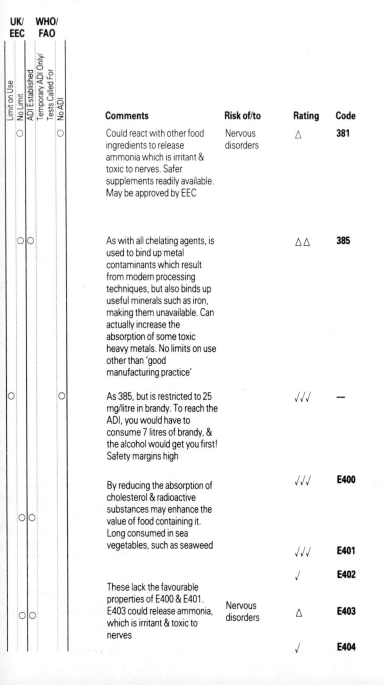

Limit on Use	No Limit	ADI Established	Temporary ADI Only/ Tests Called For	No ADI	Comments	Risk of/to	Rating	Code
	○			○	Could react with other food ingredients to release ammonia which is irritant & toxic to nerves. Safer supplements readily available. May be approved by EEC	Nervous disorders	△	**381**
	○	○			As with all chelating agents, is used to bind up metal contaminants which result from modern processing techniques, but also binds up useful minerals such as iron, making them unavailable. Can actually increase the absorption of some toxic heavy metals. No limits on use other than 'good manufacturing practice'		△△	**385**
○				○	As 385, but is restricted to 25 mg/litre in brandy. To reach the ADI, you would have to consume 7 litres of brandy, & the alcohol would get you first! Safety margins high		√√√	—
	○	○			By reducing the absorption of cholesterol & radioactive substances may enhance the value of food containing it. Long consumed in sea vegetables, such as seaweed		√√√	**E400**
							√√√	**E401**
							√	**E402**
	○	○			These lack the favourable properties of E400 & E401. E403 could release ammonia, which is irritant & toxic to nerves	Nervous disorders	△	**E403**
							√	**E404**

Emulsifiers, Stabilisers, Thickeners and Others E405–E412

Code	Names	Source	Uses	Examples
E405	Propane-1,2-diol alginate, Propylene glycol alginate, Alginate ester	Solvent Ester	As E402–4	As E402–4
E406	Agar, Agar-agar, Japanese isinglass	Seaweed derivative	Thickener, stabiliser, gelling agent	Frozen gateaux, ice-cream, glazed meat products
E407	Carrageenan, Irish moss	Seaweed extract	Emulsifier, thickener, gelling agent	Wide range of foods: cakes, ca mixes, biscuits, ice-cream, mousses, jelly crystals, quick-s jells, cheese spreads, tinned turkey breast. Permitted in bab foods
E410	Locust bean gum, Carob gum	Locust beans	Emulsifier, thickener, gelling agent	Yoghurt, cottage cheese, chee spreads, ice-cream, tinned Irisl stew, prepared fish dishes, mousses, jelly crystals, thousa island dressing
E412	Guar gum, Jaguar gum, Cluster bean, Guar flour	Seeds of legume cyamopsis	Emulsifier, dietetic food	Yoghurt, cottage cheese, chee spreads, prepared fish dishes, sausages, tinned meat dishes, frozen dinners, Scotch eggs, tinned lobster soup, cakes, meringue mixes, ice-cream, packet sauces, stuffed olives, coleslaw, tomato ketchup, relishes, horseradish sauce, piccalilli, pimento paste, fruity sauces

Restrictions on Use

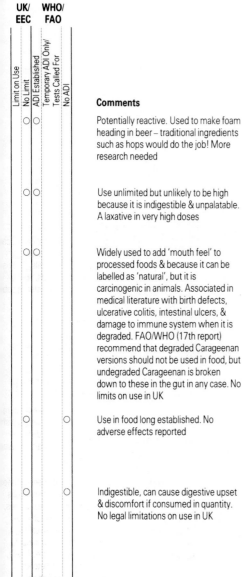

UK/EEC			WHO/FAO		Comments	Risk of/to	Rating	Code
Limit on Use	No Limit	ADI Established	Temporary ADI Only/ Tests Called For	No ADI				
	○	○			Potentially reactive. Used to make foam heading in beer – traditional ingredients such as hops would do the job! More research needed		△	**E405**
	○	○			Use unlimited but unlikely to be high because it is indigestible & unpalatable. A laxative in very high doses		√√	**E406**
	○	○			Widely used to add 'mouth feel' to processed foods & because it can be labelled as 'natural', but it is carcinogenic in animals. Associated in medical literature with birth defects, ulcerative colitis, intestinal ulcers, & damage to immune system when it is degraded. FAO/WHO (17th report) recommend that degraded Carageenan versions should not be used in food, but undegraded Carageenan is broken down to these in the gut in any case. No limits on use in UK	Conception, ?cancer, bowel disorders	△△△	**E407**
	○			○	Use in food long established. No adverse effects reported		√√√	**E410**
	○			○	Indigestible, can cause digestive upset & discomfort if consumed in quantity. No legal limitations on use in UK	Bowel disorders	△	**E412**

Emulsifiers, Stabilisers, Thickeners and Others E413–E420(ii)

Code	Names	Source	Uses	Examples
E413	Tragacanth, Gum tragacanth, Gum dragon	From wood of leguminous tree	Emulsifier, stabiliser, thickener	Frozen gateaux, salad cream, piccalilli, cheese spreads, cottage/cream cheese, sweets, diet pills
E414	Gum arabic, Gum acacia, Sudan gum, Gum hashab, Kordafan gum	From wood of acacia species	Emulsifier, stabiliser, glaze, retardant of sugar crystallisation	Tinned goulash, Irish stew, dry roasted peanuts, diet cola drinks, bitter lemon, orangeade, sweets, diet pills
E415	Xanthan gum, Corn sugar gum	Bacterial fermentation of a carbohydrate	Emulsifier, stabiliser, thickener	Cottage cheese, soft cheese, frozen dinners, cakes, cake mixes, pizzas, dessert topping mixes, sorbet mixes, chocolate flavour syrup, coleslaw, barbecue sauce, horseradish sauce, blue cheese dressing, salad cream
416	Karaya gum, Sterculia gum	Wood of SE Asian tree	Emulsifier, thickener, stabiliser, bowel regulator, diet-aid	Soft cheese, piccalilli, medicines
E420(i)	Sorbitol	Occur naturally, but synthesised commercially	Stabiliser, sweetener, glycerol, substitute	Cakes, mints, ice-cream, diabetic foods
E420(ii)	Sorbitol syrup			

	UK/EEC		WHO/FAO				Comments	Risk of/to	Rating	Code
Limit on Use	No Limit	ADI Established	Temporary ADI Only	Tests Called For	No ADI					
	○				○		ADI could not be established despite extra evidence submitted to FAO/WHO. Affects liver in animal tests. Rare allergic reactions, sometimes severe, if inhaled or consumed pure	Liver disorders	△	**E413**
	○				○		Hypersensitivity reported, but rarely & after indigestion or inhalation. Generally used at under 1% but up to 45% has been achieved in some confectionery. Use not limited. Further safety information called for by FAO/WHO	Hypersensitivity	△	**E414**
	○				○		No limits on use. No adverse effects reported		√√	**E415**
	○			○			No limits on use in UK. ADI could not be determined by FAO/WHO. Not yet approved by EEC		√	**416**
	○			○			No limits on use in UK. Could be easy, especially for diabetics, to overconsume. In animals concentrations of over 20% produced diarrhoea, flatulence, bloating. Further safety tests were required by FAO/WHO (22nd report). Banned in foods for babies. See general warning on flavour enhancers & sweeteners	Bowel disorders; diabetics	△△	**E420(i)**
	○			○					△△	**E420(ii)**

Emulsifiers, Stabilisers, Thickeners and Others E421–431

Code	Names	Source	Uses	Examples
E421	Mannitol, Manna sugar	From seaweed & coniferous wood	Sweetener, texturiser	Chewing gum, sweets, ice-crea
E422	Glycerol	Widespread in plants, by-product of soap-making	Solvent, sweetener	Chewing gum, sweets, icing, liqueurs, & as a solvent for othe additives
430	Polyoxyethylene (8) stearate, Polyoxyl (8) stearate	Synthesised	Emulsifier, stabiliser	Cakes
431	Polyoxyethylene (40) stearate, Polyoxyl (40) stearate			

Restrictions on Use

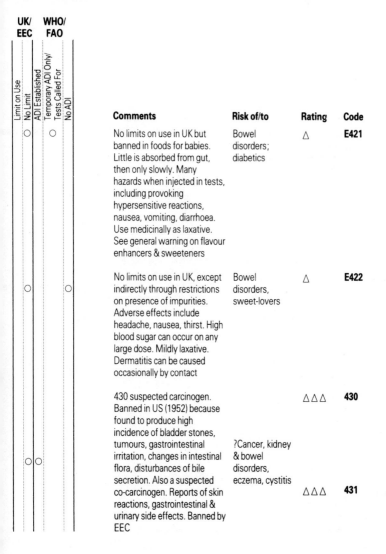

UK/EEC			WHO/FAO		Comments	Risk of/to	Rating	Code
Limit on Use	No Limit	ADI Established	Temporary ADI Only/Tests Called For	No ADI				
	O		O		No limits on use in UK but banned in foods for babies. Little is absorbed from gut, then only slowly. Many hazards when injected in tests, including provoking hypersensitive reactions, nausea, vomiting, diarrhoea. Use medicinally as laxative. See general warning on flavour enhancers & sweeteners	Bowel disorders; diabetics	△	**E421**
O				O	No limits on use in UK, except indirectly through restrictions on presence of impurities. Adverse effects include headache, nausea, thirst. High blood sugar can occur on any large dose. Mildly laxative. Dermatitis can be caused occasionally by contact	Bowel disorders, sweet-lovers	△	**E422**
		O			430 suspected carcinogen. Banned in US (1952) because found to produce high incidence of bladder stones, tumours, gastrointestinal irritation, changes in intestinal flora, disturbances of bile secretion. Also a suspected co-carcinogen. Reports of skin reactions, gastrointestinal & urinary side effects. Banned by EEC	?Cancer, kidney & bowel disorders, eczema, cystitis	△△△	**430**
		O					△△△	**431**

Code	Names	Source	Uses	Examples
432	Polyoxyethylene sorbitan monolaurate, Polysorbate 20, Tween 20			
433	Polyoxyethylene sorbitan mono-oleate, Polysorbate 80, Tween 80			
434	Polyoxyethylene sorbitan monopalmitate, Polysorbate 40, Tween 40	From sorbitol (E420)	Oil-in-water, emulsifier, stabiliser	Cakes
435	Polyoxyethylene sorbitan monostearate, Polysorbate 60, Tween 60			
436	Polyoxyethylene sorbitan tristearate, Polysorbate 65, Tween 65			
E440(a)	Pectin; Ammonium pectate; Potassium pectate; Sodium pectate	Prepared commercially from apple & orange residues	Emulsifying agent, gelling agent	Yoghurts, jams, marmalades, sauces, cakes, tarts, biscuits
E440(b)	Amidated pectin; Pectin extract			
442	Ammonium phosphates, Emulsifier YN	Synthetic, similar to lecithin	Stabiliser, emulsifier	Chocolate drops, chocolate bars, cocoa

Restrictions on Use

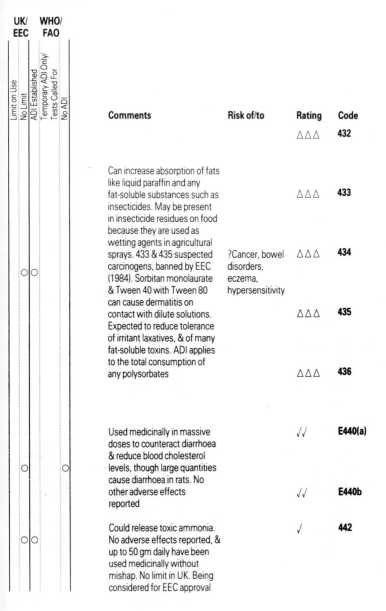

Limit on Use	No Limit	ADI Established	Temporary ADI Only	Tests Called For	No ADI	Comments	Risk of/to	Rating	Code
								△△△	**432**
						Can increase absorption of fats like liquid paraffin and any fat-soluble substances such as insecticides. May be present in insecticide residues on food because they are used as wetting agents in agricultural sprays. 433 & 435 suspected carcinogens, banned by EEC (1984). Sorbitan monolaurate & Tween 40 with Tween 80 can cause dermatitis on contact with dilute solutions. Expected to reduce tolerance of irritant laxatives, & of many fat-soluble toxins. ADI applies to the total consumption of any polysorbates		△△△	**433**
	○	○					?Cancer, bowel disorders, eczema, hypersensitivity	△△△	**434**
								△△△	**435**
								△△△	**436**
	○				○	Used medicinally in massive doses to counteract diarrhoea & reduce blood cholesterol levels, though large quantities cause diarrhoea in rats. No other adverse effects reported		√√	**E440(a)**
								√√	**E440b**
	○	○				Could release toxic ammonia. No adverse effects reported, & up to 50 gm daily have been used medicinally without mishap. No limit in UK. Being considered for EEC approval		√	**442**

Emulsifiers, Stabilisers, Thickeners and Others
E450(a)–E450(c)

Code	Names	Source	Uses	Examples
E450(a)	Disodium dihydrogen diphosphate, Acid sodium pyrophosphate; Trisodium diphosphate; Tetrasodium diphosphate, Tetrasodium pyrophosphate	Salts of phosphoric acid	Buffer, emulsifier, raising agent, sequestrant, chelating agent	Self-raising flour, baking powder, frozen breaded cauliflower, processed potatoes, fish fingers, frozen fish dinners, wide range of meat & meat products: hamburgers, salt beef, sausages, hot dogs, paté, luncheon meat, ham, bacon, gammon steaks, oven ready chickens, chicken roll, turkey burgers, chopped & shaped meat, cakes, baked apple roll, crumpets, cake mixes, sweet & savoury cheese biscuits, tinned lobster soup, packet soups, cheesecake mixes, packet pudding mixes, custard type fillings
E450(b)	Pentasodium triphosphate, Sodium tripolyphosphate; Pentapotassium triphosphate, Potassium tripolyphosphate	Synthetic	Texturiser, emulsifier, stabiliser, sequestrant. They enable up to 20% extra water to be incorporated	Cheese spreads, processed cheese slices, smoked cheese, fish fingers, wide range of meat products: hamburgers, salt beef, sausages, hot dogs, paté, luncheon meat, ham, bacon, gammon steaks, oven ready chickens, chicken roll, turkey burgers, chopped & shaped meat, cakes, baked apple roll, cake mixes, iced biscuits, sweet & savoury cheese biscuits, packet soups, tinned rice pudding
E450(c)	Sodium polyphosphates; Potassium polyphosphates			

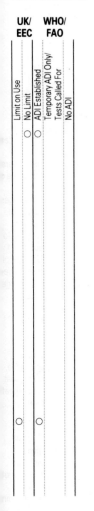

	UK/ EEC		WHO/ FAO		
Limit on Use	No Limit	ADI Established	Temporary ADI Only/ Tests Called For	No ADI	

Comments	Risk of/to	Rating	Code
Used to turn liquid stews & mixes solid, making them seem more substantial than they are. Purgative & mildly diuretic at doses well within the ADI (FAO/WHO 1974). See warnings for E338 & 385	Bowel disorders	△△△	**E450(a)**
		△△△	**E450(b)**
Added to meat to make it take up water, so increasing the weight & price without extra expense to the manufacturer. Practice is very common, can be further abused by unscrupulous manufacturers. Can interfere with digestive enzymes. See warnings for E338 & 385		△△△	**E450(c)**

Emulsifiers, Stabilisers, Thickeners and Others E460(i)–E466

Code	Names	Source	Uses	Examples
E460(i)	Microcrystalline cellulose	Chemically treated plant cellulose	Suspending & bulking agent. Wide range of applications	Much used in high-fibre or low calorie foods, also as a carrier of other additives (especially colourings & flavourings)
E460(ii)	Alpha-cellulose, Powdered cellulose	Mechanically treated plant cellulose	Bulking agent, stabiliser, suspending agent	Slimming & diet foods
E461	Methyl cellulose, Methogel, Cologel	Wood pulp		
E463	Hydroxypropyl cellulose	Treated cellulose	Bulking agent, emulsifier, stabiliser, thickener, gelling agent	Potato waffles, frozen potato shapes, hot dogs, burgers, fish fingers, Scotch eggs, flan cases, Swiss rolls, sponge cakes, ice-cream, ice-cream cones, ice-pops, mousse mixes, dessert toppings, jelly crystals, meringue, horseradish sauce, instant hot chocolate drink, range of soft drinks: bitter lemon, tonic water, orange & lemon drinks, juices in a box
E464	Hydroxypropyl methyl cellulose, Hypromellose			
E465	Ethyl methyl cellulose, Methyl ethyl cellulose			
E466	Carboxymethyl cellulose sodium salt, Carmellose sodium, CMC			

Restrictions on Use

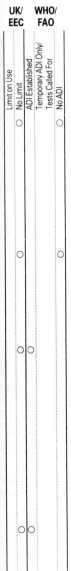

UK/EEC			WHO/FAO				
Limit on Use	No Limit	ADI Established	Temporary ADI Only/Tests Called For	No ADI	Comments	Risk of/to	Rating	Code

	O			O	Used to bulk out slimming foods. No adverse effects reported in humans & those seen in long-term animal studies were put down to deficiencies in the high bulk, low nutrient diet. Of course, humans could suffer in the same way. We should not assume that the effects on digestion will be harmless just because they are found in natural plant fibre. Banned in UK in foods for babies	Bowel disorders; anyone underweight	△	**E461(i)**
	O			O	No adverse effects reported; less likely to have unpredictable effects than chemically altered variety E460(i). Not permitted in UK in foods for babies. Can assume it will stimulate bowel activity, & dilute nutrients – giving less value for given bulk	Bowel disorders; anyone underweight	△	**E460(ii)**
	O	O					△	**E461**
							△	**E463**
					E466 the most commonly used and a suspected carcinogen. Digestive discomfort, flatulence, bloating in medicinal use at doses comparable to those permitted at WHO/FAO ADI. This may well be exceeded by those on diets, especially as it applies to the total of modified celluloses. Intestinal obstruction a theoretical risk		△	**E464**
						?Cancer, bowel disorders		
	O	O					△	**E465**
							△ △ △	**E466**

Emulsifiers, Stabilisers, Thickeners and Others E470–E472(c)

Code	Names	Source	Uses	Examples
E470	Salts of fatty acids, Soaps	Synthetic	Emulsifiers, stabilisers	Dutch type rusks. Swiss rolls, malted barley, crisps
E471	Glycerides of fatty acids; Glyceryl monostearate; Glyceryl distearate	Synthetic	Emulsifiers, solvents, stabilisers	Breads: brown, white, wholemeal, burger buns & rolls, bread mixes & as carrier for other additives in bakery. Dried skimmed milk, margarine, low fat spreads, instant mash potato, frozen vegetables, pork pies, fish products, wide range of cakes, flan cases, Swiss rolls, malt loaf, doughnuts, cake mixes, frozen pastry, sweet & savoury biscuits, liquorice-all-sorts, ice-cream, instant custard powder, custard type fillings, instant & packet soups, tinned lobster soup, instant sauces, cheese sauce mix, savoury rice, frozen pizza, mousses, popcorn, savoury snacks, instant hot chocolate drink, coffee whitener, peanut butter, cosmetics
E472(a)	Acetic acid esters of glycerides of fatty acids, Acetoglycerides, Glycerol esters			Bread: brown, white & wholemeal, rolls, croissants, burger buns, bread mixes, range of cakes, sponges, flans & tarts, cake mixes, sweet & savoury biscuits, ice-cream, instant soups, tinned mushroom soup, frozen pizza, savoury rice, gravy granules, cheese sauce granules & mix, cheesecake mix, dessert topping mix, instant hot chocolate drink
E472(b)	Lactic acid esters of glycerides of fatty acids, Lactylated glycerides, Lactoglycerides	Synthetic	Emulsifiers, stabilisers	
E472(c)	Citric acid esters of glycerides of fatty acids			

Restrictions on Use

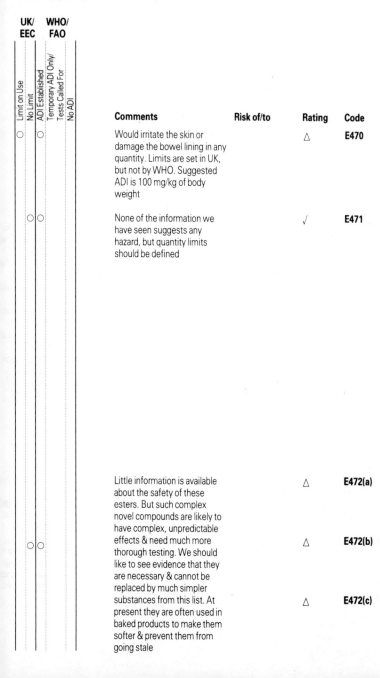

UK/EEC			WHO/FAO		Comments	Risk of/to	Rating	Code
Limit on Use	No Limit	ADI Established	Temporary ADI Only/Tests Called For	No ADI				
○		○			Would irritate the skin or damage the bowel lining in any quantity. Limits are set in UK, but not by WHO. Suggested ADI is 100 mg/kg of body weight		△	**E470**
	○	○			None of the information we have seen suggests any hazard, but quantity limits should be defined		√	**E471**
					Little information is available about the safety of these esters. But such complex novel compounds are likely to have complex, unpredictable effects & need much more thorough testing. We should like to see evidence that they are necessary & cannot be replaced by much simpler substances from this list. At present they are often used in baked products to make them softer & prevent them from going stale		△	**E472(a)**
	○	○					△	**E472(b)**
							△	**E472(c)**

Code	Names	Source	Uses	Examples
E472(d)	Tartaric acid esters of glycerides of fatty acids			
E472(e)	Mono and diacetyltartaric acid esters of glycerides of fatty acids	As E472(a)–(c)		
E473	Sucrose esters of fatty acids			
E474	Sucroglycerides			
E475	Polyglycerol esters of fatty acids			
476	Polyglycerol esters of polycondensed fatty acids of castor oil, Polyglycerol polyricinoleate	From castor oil	Emulsifier, stabiliser	Chocolate biscuits, chocolate flavour coatings, soya bean & greasing emulsions for baking tins
—	Polyglycerol esters of dimerised fatty acids of soya bean oil	From soya beans	Emulsifier, stabiliser	
E477	Propylene glycol esters of fatty acids, Propane-1,2-diol esters of fatty acids	From propylene glycol	Emulsifier, stabiliser	Cake mixes, cheesecake mix, packet pudding mixes, dessert topping mixes, mousses. Carri for added flavourings
478	Lactylated fatty acid esters of glycerol and propane-1,2-diol	From lactic acid	Emulsifier, stabiliser, surfactant	

Restrictions on Use

UK/EEC		WHO/FAO						
Limit on Use	No Limit	ADI Established	Temporary ADI Only/Tests Called For	No ADI	Comments	Risk of/to	Rating	Code
							△	**E472(d)**
							△	**E472(e)**
	○	○						
							△	**E473**
							△	**E474**
							△	**E475**
○	○				Being considered for acceptance by EEC. Derived from a highly irritant oil whose use is limited to tiny amounts on its own in UK & by WHO. Seems a heavy sledgehammer to crack a tiny nut, when the list of alternatives is so long		△△	**476**
					No information available		—	
	○	○			Based on fairly reactive substance, which is strongly sensitising in skin creams. Too little information for confidence		△	**E477**
	○			○	Propane-l, 2-diol is strongly sensitising in skin creams. Too little information available for confident use		△	**478**

Emulsifiers, Stabilisers, Thickeners and Others E481–495

Code	Names	Source	Uses	Examples
E481	Sodium stearoyl-2-lactylate	Lactic acid derivatives	Emulsifier, stabiliser	Bread, burger buns, melba to biscuits, frozen pizza, frozen dinners, gravy granules
E482	Calcium stearoyl-2-lactylate			
E483	Stearyl tartrate	Tartaric acid derivative	Emulsifier, stabiliser	Bread
491	Sorbitan monostearate	Synthetic from sorbitol	Water-in-oil emulsifiers, stabilisers, carrier for other additives	Dried yeast, cream fillings. Ca for other additives
492	Sorbitan tristearate, Span 65			
493	Sorbitan monolaurate, Span 20			
494	Sorbitan mono-oleate, Span 80			
495	Sorbitan monopalmitate, Span 40			

Restrictions on Use

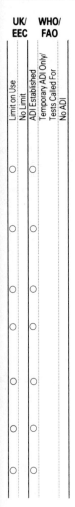

UK/EEC		WHO/FAO			Comments	Risk of/to	Rating	Code
Limit on Use	No Limit	ADI Established	Temporary ADI Only/Tests Called For	No ADI				
							√	**E481**
○		○			Little information found to justify inclusion in a long list of substances more obviously safe		√	**E482**
○		○			Little information found. Permitted in high quantities. Arouses weaker misgivings than its neighbours in this list		√	**E483**
○		○					△△	**491**
○		○					△△	**492**
○		○			Similar to but simpler than 432–6, 493, a suspected carcinogen. More reactive & more likely to be absorbed, but with different physical properties. Dilute solutions of Span 20 can cause dermatitis on contact, as can mixtures of Span 60 & Span 80	?Cancer, bowel disorders, hypersensitivity, eczema	△△△	**493**
○		○					△△	**494**
○		○					△△	**495**

Acids, Bases and Others (500–529)

Acids are used to give a tart flavour to food, and as raising agents or preservatives. Bases increase the alkalinity of food.

Many of the additives in this group are processing aids and so do not need to be declared on labels. They may be additives for additives – for example, 527 ammonium hydroxide is a solvent for colourings; or they may be additives made necessary by modern processing techniques – for example, metal machinery leaves traces of metal contaminants in food which could be dangerous, so sequestrants or chelating agents are added to trap the metals and make them harmless. Unfortunately, they may at the same time trap vital minerals in the food, such as iron, and prevent them being used as nutrients in the body.

Code	Names	Source	Uses	Examples
500	Sodium carbonate; Sodium hydrogen carbonate, Sodium bicarbonate, Baking soda, Bicarbonate of soda; Sodium sesquicarbonate Trona	Industrial manufacture	Raising agent, buffer, base	Self-raising flour, baking powd crumpets & other bakery products, dried potatoes, froze breaded vegetables, cake mix biscuits, sweets, ice-cream cones, tinned rice pudding, custard powder, fizzy tablets
501	Potassium carbonate; Potassium hydrogen carbonate			
503	Ammonium carbonate; Ammonium hydrogen carbonate, Ammonium bicarbonate, Hartshorn	Manufacture	Buffer, aerating agent, base	Custard cream biscuits
504	Magnesium carbonate, Magnesite	Naturally occurring mineral	Base, buffer, anti-caking agent	Bread, salt, ice-cream cones, icing sugar
507	Hydrochloric acid	Manufacture	Acid	

Restrictions on Use

UK/EEC		WHO/FAO				Comments	Risk of/to	Rating	Code
Limit on Use	No Limit	ADI Established	Temporary ADI Only	Tests Called For	No ADI				
							Bowel disorders	√	**500**
○		○				Long-established; used medicinally in moderate doses as antacid. Potential irritant of the bowel, & in large quantities could disturb the acid-base balance of the blood. Awaiting EEC approval. No ADI has been established & they are used in fair quantity		√	**501**
○		○				Could release ammonia, which is irritant & toxic to nerves, on decomposition. Used in sufficient quantity for this to seem a real risk, despite WHO/FAO's confidence. Awaiting EEC approval	Nervous disorders	△	**503**
○		○				Not very soluble in water, so less reactive than 500–503. Has nutritional value, but can cause diarrhoea. Awaiting EEC approval		√	**504**
	○	○				Although this occurs naturally in stomach juice, it is irritant & corrosive. It is absurd that no limit has been set on the use of such a strong, reactive acid in food, which has to pass through the mouth & throat too. Awaiting EEC approval	Irritation to stomach and mouth	△	**507**

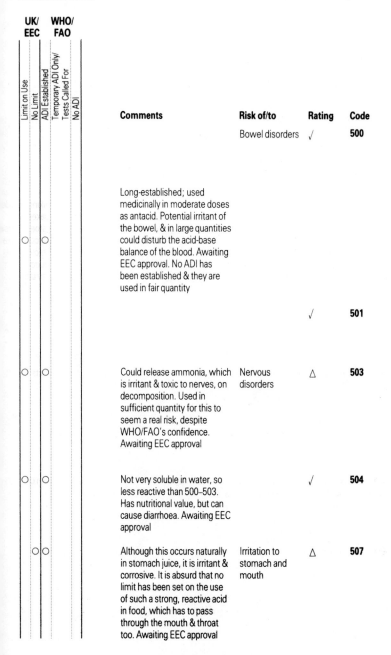

Code	Names	Source	Uses	Examples
508	Potassium chloride	Naturally occurring in salt deposits	Salt substitute, nutrient	Low salt condiments, jelly crystals, quick-set jells, frozen vegetables
509	Calcium chloride	Contained in salt deposits, & easily manufactured	Sequestrant, firming agent	Tinned fruit & vegetables, frozen vegetables, pimento paste
510	Ammonium chloride	Manufactured	Nutrient	Yeast culture nutrient
513	Sulphuric acid	Manufactured	Acid	
514	Sodium sulphate	Naturally occurring mineral	Diluent	Frozen vegetables, low salt condiment
515	Potassium sulphate	Naturally occurring mineral	Salt substitute	Frozen vegetables, low salt condiment
516	Calcium sulphate, Gypsum, Plaster of Paris	Naturally occurring mineral	Excipient, firming agent, nutrient, sequestrant	Bread, frozen vegetables, pills, yeast culture nutrient

Restrictions on Use

UK/EEC		WHO/FAO			Comments	Risk to/of	Rating	Code
Limit on Use	No Limit	ADI Established	Temporary ADI Only/Tests Called For	No ADI				
	○			○	Produces nausea in some people. When used as a medicinal supplement, causes gastric & intestinal erosions which can be severe. The range of food & limitations on use should be published & kept small	Irritation to stomach	△	**508**
○	○				Irritant to the stomach. Should therefore be subject to restrictions, but currently no limits on use in food. Being considered for EEC approval	Irritation to stomach	△	**509**
	○	○			Another substance whose unlimited use should not be sanctioned. Readily absorbed & could release ammonia which is toxic to the nerves into the bloodstream or liver. Foods in which it is allowed & limits should be stipulated. Awaiting EEC approval	Liver disease, nervous disorders	△	**510**
○				○	A strong, reactive acid; unquestionably poisonous & irritant. Suspected carcinogen. Limits should be defined, if any can be justified	?Cancer	△△△	**513**
○				○	No limits are set. Given its function, quantities used might be appreciable. Limits should be stated	Babies and young children	△	**514**
	○			○	Mildly purgative in gram doses, but harmless otherwise to those with healthy kidneys who can cope with potassium stress		√	**515**
	○	○			Inert & insoluble in food, but partly dissolves in stomach acid. No hazards known. Awaiting EEC approval		√	**516**

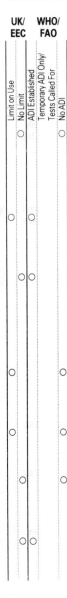

Code	Names	Source	Uses	Examples
—	Ammonium sulphate	Manufacture		Frozen vegetables
518	Magnesium sulphate	Naturally occurring mineral	Firming agent	Epsom salts, beer
524	Sodium hydroxide	Synthetic	Base	Savoury biscuits
525	Potassium hydroxide			
526	Calcium hydroxide	Processing of natural lime	Base, firming agent	Tinned fruit & vegetables, hard cheese
527	Ammonium hydroxide	Synthetic	Solvent for colourings, base	Used in other additives e.g. colourings
528	Magnesium hydroxide	Processing of natural rocks	Base	
529	Calcium oxide	Processing of natural rock	Base	

Restrictions on Use

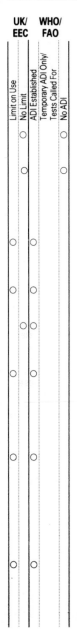

UK/EEC			WHO/FAO			Comments	Risk of/to	Rating	Code
Limit on Use	No Limit	ADI Established	Temporary ADI Only	Tests Called For	No ADI				
	○				○	Could release ammonia which is neurotoxic		△	—
	○				○	A valuable nutrient. Although a competitor with zinc, iron & calcium for absorption from the gut, unlikely to cause problems in normal food use		√√	**518**
○	○					Highly toxic, irritant & reactive, so that exposure to any appreciable amount would be hazardous. Should be subject to strict, explicit controls. Awaiting EEC approval		△△	**524**
○	○							△△	**525**
	○	○				A weak base, much less hazardous than other hydroxides (524, 525, 527)		√√	**526**
○	○					Caustic & irritant, because strongly alkaline. Can also release ammonia, which is neurotoxic, once absorbed. Uses should be strictly controlled	Nervous disorders	△△	**527**
○	○					Less hazardous than other metal hydroxides, but can cause diarrhoea. Might reduce the effectiveness of ulcer treatment. Increases the absorption of calcium & magnesium from foods in gram amounts		√	**528**
○	○					On contact with moisture would form calcium hydroxide (526). Rather reactive for food use in our opinion. Food uses should be justified before they are sanctioned		√	**529**

Anti-caking Agents and Others (530–578)

Anti-caking agents are chemicals added to powdered or granular foods to prevent them absorbing moisture and sticking together. Many of the additives in this group serve other purposes as well. They may be excipients – additive powders used as 'carriers' of other additives, or dusting powders and releasing agents – used to prevent processed foods sticking to manufacturing plant. Despite the long list of alternatives in this group, those whose safety is most in doubt are among the most commonly used.

Code	Names	Source	Uses	Examples
530	Magnesium oxide, Native magnesium, Periclase	Mineral occurs naturally	Anti-caking agent, base	
535	Sodium ferrocyanide, Sodium hexacyano-ferrate II	Synthesised	Anti-caking agent	Table salt
536	Potassium ferrocyanide, Potassium hexacyano-ferrate II			
540	Dicalcium diphosphate, Calcium hydrogen phosphate	Mineral or manufactured	Buffer, nutrient, abrasive, excipient	Toothpaste, pills, supplements, cheese
541	Sodium aluminium phosphate acidic; Sodium aluminium phosphate basic	Manufactured	Raising agent, emulsifier	Frozen breaded vegetables
542	Edible bone phosphate	Animal bones	Anti-caking agent, nutrient, excipient	Mineral supplement, pills

Restrictions on Use

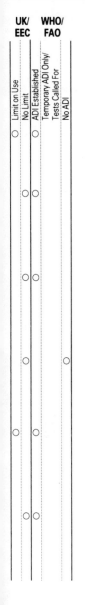

	UK/EEC		WHO/FAO			Comments	Risk of/to	Rating	Code
Limit on Use	No Limit	ADI Established	Temporary ADI Only/Tests Called For	No ADI				
O		O			Mildly laxative, but its properties depend on its physical form. No reports of toxicity if inhaled in small amounts. Awaiting EEC approval		√	**530**
	O	O					△ △ △	**535**
	O	O			Said to be safe because the cyanide in these compounds is strongly bonded & inert, but WHO/FAO have set a very low ADI. In anti-caking agents we do not think this kind of risk is worth taking		△ △ △	**536**
	O			O	Does not dissolve in water, but does dissolve in stomach acid. Used medicinally up to 2 gm daily without ill effect. Awaiting EEC approval		√√√	**540**
O	O				Likely to release some aluminium during acid digestion, which is undesirable however slight		△	**541**
	O	O			Vulnerable to the same trace contaminants as the animals from which it comes. Vegetarians will wish to avoid it. Harmless otherwise. Awaiting EEC approval		√	**542**

Code	Names	Source	Uses	Examples
544	Calcium polyphosphates	⎫ Manufactured	Emulsifiers, water-retainers, chelating agents	Packaged meat, frozen chicken, cheese
545	Ammonium polyphosphates	⎭		
551	Silicon dioxide; Silicea; Silica	From sand by mechanical processing & hydration	Stabiliser, anti-caking agent, suspending agent	Sweets, crisps
552	Calcium silicate	Synthetic.	Anti-caking agent, glaze, polish, dusting, release agent	Rice, salt, sweets, icing sugar
553(a)	Magnesium silicate synthetic; Magnesium trisilicate	Manufactured Naturally occurring rock	Anti-caking agent, glaze, polish, dusting, excipient	Rice, salt, tinned foods, pills, sweets, icing sugar
553(b)	Talc	Naturally occurring mineral	Release agent, lubricant	As 553(a)

Restrictions on Use

	UK/EEC		WHO/FAO		Comments	Risk of/to	Rating	Code
Limit on Use	No Limit	ADI Established	Temporary ADI Only/Tests Called For	No ADI				
	O	O			May interfere with digestive enzymes. As chelating agents may also prevent absorption of vital minerals such as iron, while increasing absorption of some toxic minerals such as lead. 545 may release toxic ammonia. Can be used to reduce the value of expensive items by adding water. Require caution		△	**544**
							△	**545**
	O	O			Yes, they actually put sand in our food. Use in foods limited only by good manufacturing practice. Gel forms said to be safe, but six workers got silicosis (lung disease) after working briefly with the powder		√	**551**
	O	O			Forms a gel with stomach acid, neutralising excess acidity. No evidence of risk from inhalation.		√√	**552**
	O			O	Forms silicagel in stomach, which can neutralise digoxin (a heart drug). Harmless in itself		√√	**553(a)**
	O			O	Magnesium silicate with some aluminium silicate. Suspected carcinogen. Some sources contain traces of asbestos, but much larger quantities used cosmetically. Risk of lung disease from long-term inhalation, independent of any asbestos contact	?Cancer, lung disease	△△△	**553(b)**

Anti-caking Agents and Others 554–572

Code	Names	Source	Uses	Examples
554	Aluminium sodium silicate			
556	Aluminium calcium silicate, Calcium aluminium silicate	Natural mineral	Anti-caking agent	Coffee whitener
558	Bentonite, Bentonitum, Soap clay	A natural aluminium silicate	Stabiliser of suspensions & emulsions, wine clarifier	Wine
559	Kaolin heavy, Kaolin light	A natural aluminium silicate	Anti-caking agent	Wine
570	Stearic acid	Synthetic, but naturally widespread	Anti-caking agent, lubricant, tablet coating	Pills
—	Butyl stearate			
572	Magnesium stearate	Synthetic	Anti-caking agent, dusting powder, release agent, emulsifier, lubricant	Frozen vegetables, pills,
—	Calcium stearate			

Restrictions on Use

UK/EEC			WHO/FAO					
Limit on Use	No Limit	ADI Established	Temporary ADI Only/ Tests Called For	No ADI		Comments	Risk of/to	Rating	Code

UK/EEC			WHO/FAO		Comments	Risk of/to	Rating	Code
							△	**554**
	○	○			The small amounts of aluminium which may be absorbed are a hazard to at least some people. No reports found of risk from inhalation		△	**556**
	○			○	Forms a dissolved web which carries impurities to bottom. Harmless, unless intolerant of aluminium		√	**558**
	○			○	Lung diseases followed inhalation of kaolin in workers handling it. It is a powerful absorbent of toxins; used medicinally; wet forms in drink are harmless, as insoluble even in stomach acid		√	**559**
	○			○	No adverse effects reported from pharmaceutical experience. Scarcely dissolves in water, so remains in the gut unabsorbed		√√√	**570**
					Scanty information but similar to 570		√	—
	○	○			Deaths have occurred from inhalation of baby dusting powder, so not safe when used dry		△	**572**
					Little information found but presumed similar to 572		△	—

Code	Names	Source	Uses	Examples
575	D-glucono-1,5-lactone, Glucono deltalactone	Manufactured from glucose	Sequestrant	Milk processing, residue in bee cake mixes
576	Sodium gluconate			
577	Potassium gluconate	Manufactured from gluconic acid	Sequestrant, nutrient	
578	Calcium gluconate			

Restrictions on Use

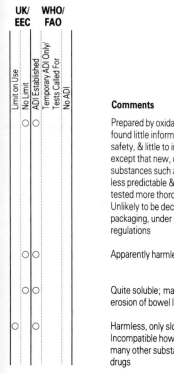

	UK/EEC		WHO/FAO					
Limit on Use	No Limit	ADI Established	Temporary ADI Only/Tests Called For	No ADI	Comments	Risk of/to	Rating	Code
	O	O			Prepared by oxidation. We found little information on safety, & little to incriminate it except that new, complex substances such as this are less predictable & should be tested more thoroughly. Unlikely to be declared on packaging, under present regulations		√	**575**
	O	O			Apparently harmless		√√√	**576**
	O	O			Quite soluble; may cause erosion of bowel lining		√	**577**
O		O			Harmless, only slowly soluble. Incompatible however with many other substances & drugs		√	**578**

Flavour Enhancers and Sweeteners (620–637, and unnumbered)

All the substances listed in this group are frequently used to make attractive highly processed foods which would otherwise be inedible. Flavour enhancers have little or no flavour in themselves but make the flavours of other foods stronger, often just by stimulating the taste buds on the tongue.

The use of sweeteners and refined sugars in nearly all processed products, whether sweet or savoury, has perverted our ability to choose instinctively foods which are good for us. We have included a list here of the refined sugars used in processing by the food industry – these are officially classed as foods rather than additives (because they provide calories), but they are of no nutritional value and they are so highly refined as to constitute pure chemical substances. Once separated from their biological context in this way, they work quite differently in the body and their effects on health are devastating. The average person in this country consumes 100 lbs of processed sugars a year. Our intake of processed sugars causes tooth decay and obesity, and has been linked with diabetes and heart disease. About 60 per cent of sugars consumed in this country are in processed foods. As fast as people cut down on packet sugar, more and more processed sugars are hidden in other products. Watch out for 'unsweetened fruit juice' which may quite legally contain up to 10 per cent sugar, so long as it is listed in the small print.

Flavour Enhancers and Sweeteners 620–635

Code	Names	Source	Uses	Examples
620	L-glutamic acid	All plant & animal tissues, manufactured bacteriologically	Flavour enhancer, salt-free condiment	Extremely common in processed foods of all sorts. Potato waffles, packet scalloped potatoes, fish fingers, fish cakes, fish in batter, scallops, tinned prawns, wide range of meat products: beefburgers, sausages, pork pies, gammon steaks, frozen chicken, savoury & cheese biscuits, instant soups, tinned soups, packet soups, wide range of convenience foods, savoury rice, pot casseroles, tinned meat dishes, tinned ravioli, beans in chilli sauce, frozen dinners, pizzas, packet sauces, instant sauces, stock cubes, stuffing mix, crisps, peanuts
621	Monosodium glutamate, Sodium hydrogen L-glutamate, MSG	Manufactured bacteriologically		
622	Potassium hydrogen L-glutamate, Monopotassium glutamate			
623	Calcium dihydrogen di-L-glutamate, Calcium glutamate			
627	Guanosine 5'-disodium phosphate, Sodium guanylate, Disodium guanylate	Manufactured	Flavour enhancer	Common flavour enhancers used in potato waffles, packet scalloped potatoes, instant, tinned & packet soups, pot casseroles, pot rice, savoury packet rice dishes, stock cubes, crisps, & in many products with MSG
631	Inosine 5'-disodium phosphate, Sodium 5'-inosinate	Prepared from animal muscle-waste		
635	Sodium 5'-ribonucleotide	Manufactured		

Restrictions on Use

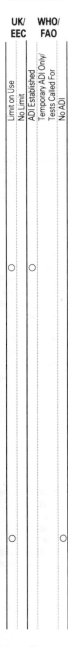

	UK/EEC		WHO/FAO		
Limit on Use	No Limit	ADI Established	Temporary ADI Only/ Tests Called For	No ADI	

Comments	Risk of/to	Rating	Code
621 is toxic to sensitised people, causing flushing, headache & chest pain (Chinese Restaurant Syndrome). But up to 12 g of 620 have been used medicinally without ill effect. The ADI sanctions up to 8.5 gm daily (½ oz) of all combined, in addition to that naturally present in food – which implies high tolerance. 621 causes cellular brain damage in newborn animals & is prohibited from foods intended for young children in UK, because of its nerve stimulant properties. Often used as a substitute for expensive ingredients & ironically to enhance the flavours, not of real foods, but of synthesised flavourings. See general warning at beginning of section	Hypersensitive. Babies & young children	√	**620**
		△ △ △	**621**
		√	**622**
		√	**623**
		△	**627**
No reports found of adverse effects from either – 635 is a mixture of the other two. An ADI is not specified, but they are banned in UK from foods intended for young children. They are converted to uric acid, so could cause problems for gout sufferers. Used to cut down on expensive ingredients, often combined with MSG. See general warning at beginning of section	Babies & young children, gout sufferers	△	**631**
		△	**635**

(○ in No Limit column under UK/EEC and under WHO/FAO for the 620–627 group; ○ in No Limit under UK/EEC and ○ under WHO/FAO No ADI for the 631–635 group)

Code	Names	Source	Uses	Examples
636	Maltol	Available in wood & roasted malt, but manufactured chemically	Flavour	636 used to give fresh baked smell to hot bread, also in cake, fruit, vanilla & chocolate flavouring & drinks
637	Ethyl maltol	Manufactured	Sweetener	
—	Saccharin			
—	Calcium saccharin			
—	Sodium saccharin			
—	Aspartame			
—	Acesulfame potassium			
—	Isomalt		Sweetener	
—	Xylitol			
—	Hydrogenated glucose syrup			
—	Thaumatin			
—	Mannitol (E421)			
—	Sorbitol (E420)			
—	Sorbitol syrup (E420)			
—	Glycerol (E422)			
—	Sucrose Sugar (Dried) glucose (syrup)			
	Dextrose	Refined from sugar beet or cane	Sweetener, to add 'mouth feel'	
	Invert sugar Fructose Lactose Maltose			

Restrictions on Use

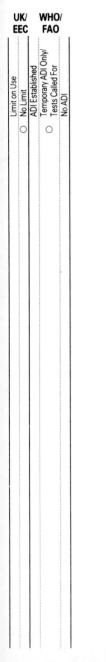

UK/EEC			WHO/FAO		Comments	Risks of/to	Rating	Code
Limit on Use	No Limit	ADI Established	Temporary ADI Only/ Tests Called For	No ADI				
	○		○		Little information available, & not obviously safe. Often used as a substitute for expensive ingredients & ironically to enhance the flavours not of real foods but of synthesised flavourings		△	**636**
							△	**637**
					All these items, described more fully elsewhere, are listed here for inclusion in the general class of sweeteners permitted in Britain. The general warning at the beginning of the section applies to all. See also section on unnumbered additives below		△△△	
					Officially considered foods rather than additives, but these are so highly refined as to constitute pure chemical substances & to have lost their biological nature as part of foods. Their use in large quantities in a wide range of processed foods, whether sweet or savoury, has devastating effects on health. High consumption of refined sugars causes tooth decay, obesity & is linked with diabetes & heart disease. They certainly come under the general warning of this section		△△△	

Glazing Agents (900–907)

This group covers waxes and oils, such as liquid paraffin, which are used to put a shine on foods like dried fruit and sweets. Many EEC countries prohibit their use and they do not have E numbers. There are no restrictions on the levels at which they may be used (with the exception of carnauba wax). The mineral hydrocarbons (905, made from petrol) are particularly suspect. They mainly occur in foods which are unlikely to be labelled or on raw ingredients for other products, e.g. raisins for muesli, where they are once again unlikely to be declared.

Glazing Agents 900–907

Code	Names	Source	Uses	Examples
900	Dimethyl polysiloxane, Simethicone, Dimethicone	Manufactured	Water-repellant, anti-foaming agent	Frozen vegetables, glucose syrup, pineapple juice
901	Beeswax, white; Beeswax, yellow	Honeycomb	Glazing & polishing agent, releasing agent, solvent for colourings and glazes in confectionery	Solvents for colourings & glazes in confectionery
903	Carnauba wax	Leaves of Brazilian wax palm	Glazing agent	Chocolate products, sweets, pill
904	Shellac	Secretion of an Indian insect	Glazing agent	Pill coatings, sweets, hair spray
905	Mineral hydrocarbons	Manufactured from petroleum	Polishing & glazing agent	Chocolate sweets, chewing gum dried fruit, citrus fruit, cheese rin
907	Refined microcrystalline wax	Manufactured from petroleum	Polishing & glazing agent	

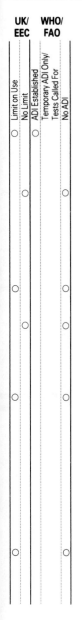

| UK/EEC | | WHO/FAO | | | | | | |
Limit on Use	No Limit	ADI Established	Temporary ADI Only/Tests Called For	No ADI	Comments	Risk of/to	Rating	Code
○		○			Suspected carcinogen. Can contain the irritant formaldehyde. Main use is to reduce foaming	?Cancer	△△△	**900**
	○			○	The white is bleached. Hypersensitivity has occasionally been reported, perhaps caused by resins in the wax		√	**901**
○				○	No reports of adverse effects found		√√√	**903**
	○			○	Hazardous to the lungs of those exposed to inhaling it occupationally – e.g. hairdressers. Otherwise harmless		√	**904**
○				○	Suspected carcinogens. Anal seepage & irritation may result from heavy consumption. A proportion is absorbed & may provoke inflammatory reactions. Produce severe pneumonia when inhaled. Prolonged consumption may interfere with absorption of vitamins A, D, E, K, & should be avoided. Dangerous to manufacturing workers	?Cancer, bowel disorders	△△△	**905**
○				○	Suspected carcinogen. Insoluble and inert, but provokes tissue reactions if absorbed. Dangerous to manufacturing workers	?Cancer	△△△	**907**

Improvers and Bleaching Agents (920–927)

Most of these are banned in food in other EEC countries. Like preservatives and antioxidants, the chemical characteristics which make them useful as bleaches and improvers also make them dangerous to human health.

They are used to make flour 'whiter than white', but in doing this they also sterilise it and so enable manufacturers to keep it almost indefinitely. Sterile flour is flour stripped of its nutrients, and it is not only of little interest to weevils, but it is also of little value to humans.

These additives also speed up the maturing of flour and make bread dough easier to process. Without them most of the white bread in this country could not be made. It is produced by the high-speed Chorleywood Process, which bypasses the normal time required in breadmaking for dough to 'prove'. Manufacturers find that wholemeal flour, with all its nutrients intact, is not so easily manipulated.

These additives are restricted to use in bread and flour and some limits are set on the levels at which they may be used, but because they are processing aids, they are very rarely declared on labels.

Code	Names	Source	Uses	Examples
920	L-cysteine hydrochloride, L-cysteine hydrochloride monohydrate	Naturally occurring amino-acid. Synthesised	Improver	Refined flour & therefore bre baps, rolls (but not wholemea cakes, biscuits; not necessa declared
924	Potassium bromate	Synthetic	Bleaches & matures flour, oxidising agent	Bread, baps & rolls (not wholemeal), white bread flou cakes, biscuits; not necessar declared
925	Chlorine	Synthetic	Preservative, bleach & flour improver	Flour, bread, baps, rolls, bisc not necessarily declared
926	Chlorine dioxide			
927	Azodicarbona- mide, Azoformamide	Synthetic	Improver & maturer	Flour (but not wholemeal), bre baps, rolls, biscuits, cakes; n necessarily declared

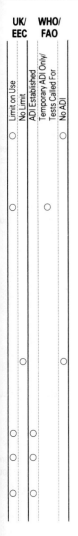

Restrictions on Use

UK/EEC — Limit on Use | No Limit | ADI Established
WHO/FAO — Temporary ADI Only/Tests Called For | No ADI

	Comments	Risk of/to	Rating	Code
Limit on Use ○ ... No ADI ○	Used medicinally to prevent ulceration of burnt eyes, & by mouth as a nutrient. Harmless		√√√	**920**
Limit on Use ○ ; Temporary ADI Only ○	A powerful, toxic oxidising agent. Has had devastating effects when consumed accidentally at 1.1% in bread (275 times the permitted maximum of 40 ppm). Would produce nausea, vomiting, abdominal pain, diarrhoea in much smaller concentrations. Destroys nutrients, especially vitamin E, in the flour it is used to bleach & mature	Bowel disorders, irritation to stomach	△△△	**924**
No Limit ○ ; No ADI ○	Very powerful bleaches & antiseptics, but very irritant & destructive. Destroys vitamin E & B$_1$, as well as oxidising other nutrients. Dangerous to workers. Banned elsewhere for use in food	Lung disease	△△△	**925**
Limit on Use ○ ; ADI Established ○			△△△	**926**
Limit on Use ○ ; ADI Established ○	Improves the processing properties of dough. We have found no reports of adverse effects, & animal tests in US suggest it is safe, but more research needed		√	**927**

Unnumbered Additives

There is a whole group of additives which have not been coded and do not appear on labels, nor in the government's guide on additives for consumers called *Look at the Label* (see bibliography, p. 276). Nevertheless, they make their way into our food. Several of them are highly suspect, and we see no reason why full lists of these additives, with details of what tests have been done on them, what they are used for and where they are likely to be found, should not be published.

Unnumbered Additives

Names	Source	Uses	Examples
Acesulfame potassium, Acesulfame K	Synthetic	Intense sweetener	Reduced sugar jams, jellies & marmalades, soft drinks, diabetic products
2-Aminoethanol, Monoethanola-mine	Manufactured	Astringent	Peeled fruit, peeled vegetables
Aspartame	Synthetic	Intense sweetener, 200 × effect of sugar, but destroyed by heat	Low calorie products & slimming foods, wide range of diet drinks
Calcium phytate, Calcium mesoinositol-hexaphosphate	Occurs naturally & manufactured	Sequestrant	Baked goods, soft drinks, processed vegetables
Dichlorodifluoro-methane	Synthetic	Refrigerant	Frozen food
Dichloro-methane	Synthetic	Solvent for caffeine & fish oils	Coffee, fish oils

Restrictions on Use

| UK/EEC | | WHO/FAO | | | Comments | Risks of/to | Rating |
Limit on Use	No Limit	ADI Established	Temporary ADI Only/Tests Called For	No ADI			
O		O			Banned in US as suspected carcinogen. An intense sweetener, chemically more complex than sugar, but simpler than aspartame. We found no information about its properties. Not permitted in foods for babies & young children	?Cancer; babies & young children	△ △ △
					Allergic reactions have been reported in high doses used medicinally & administered by injection. Not permitted in foods for babies & young children	Hypersensitivity; babies & young children	△
	O	O			Suspected carcinogen, though controversy rages over safety tests carried out in US. No limit set in UK, but not permitted in foods intended for babies & young children. Breakdown products include a nerve stimulant which enhances the stimulant effects of MSG (621). The validity of safety tests conducted by the manufacturer has been called into question recently by independent pathologists	?Cancer, nervous disorders; babies & young children	△ △ △
					Rather inert, but may be broken down by the enzyme phytase. Is probably harmless. By encouraging the production of phytase in the gut would reduce susceptibility to phytate – the substance in foods, such as wholemeal bread, which can prevent absorption of calcium		√
					Normal use in the plumbing of refrigerators! Takes 1 kg of food containing it daily to reach ADI, that could be achieved fairly easily. Almost certainly irritant		△
					Under suspicion in US, following tests in which rats developed cancer when exposed to solvent in air	?Cancer	△ △ △

Unnumbered Additives

Names	Source	Uses	Examples
Diethyl ether, Solvent ether	Synthetic	Solvent, carrier of flavourings	In other additives
Dioctyl sodium sulphosuccinate	Synthetic	Emulsifier, stabiliser, surfactant	Used in sugar refining & for washing bottles
Ethoxyquin	Synthetic	Antioxidant	Apples, pears
Ethyl alcohol, Ethanol	Brewed	Solvent	In other additives
Ethyl acetate	Synthesised	Solvent	Baked goods, jams, chewing gums
Glycerol monoacetate, Monoacetin; Glycerol diacetate, Diacetin, Glycerol triacetate, Triacetin	Synthetic	Solvent, carrier of flavourings, flavouring	In other additives, & as carrier of flavourings. Liable to be in foods processed on metal machinery
Glycine	Synthetic	Sequestrant	

Restrictions on Use

UK/EEC			WHO/FAO		Comments	Risk of/to	Rating
Limit on Use	No Limit	ADI Established	Temporary ADI Only/Tests Called For	No ADI			
O			O		Irritant to sensitive skin such as lips & mouth. Includes 20 ppm BHT (E321) as a stabiliser		△
	O	O			Inhibits normal peptic digestion in the stomach. May increase risk of liver damage from drugs consumed at the same time. Prolonged medicinal use causes diarrhoea, & can cause gastric erosions occasionally. Spermicidal. Temporary ADI of 2.5 mg/kg body weight/day was withdrawn for lack of safety evidence (WHO 1978)	Conception, bowel disorder, irritant to stomach, liver disease	△△△
O			O		Presumed from the level of its permitted use, & the fact that it is banned in foods for babies & young children, to be potent. No other safety information was found	Babies & young children	△
					Used in tiny amounts in food compared with % in brewed drinks		√
O			O		Irritant to lips & mouth, & depresses nerve function. UK permit is temporary & subject to further safety tests		△
O			O		Only allowed temporarily by FAC, further studies required		△△
	O		O		Used safely in high doses as a medicine to reduce gastric irritation		√√

Unnumbered Additives

Names	Source	Uses	Examples
Sodium heptonate	Synthetic	Sequestrant	Liable to be in foods processed on metal machinery
Calcium heptonate	Synthetic	Sequestrant	Liable to be in foods processed on metal machinery
Hydrogen	Manufactured	Packaging gas	Products in pressurised containers
Hydrogenated glucose syrup, Hydrogenated high maltose glucose syrup	Manufactured	Bulk sweetener	
Isomalt	Manufactured	Bulk sweetener	Range of sweet foods
Isopropyl alcohol	Manufactured	Solvent	As carrier for other additives
Nitrogen	Principle gas in air	Packaging gas	Foods packaged in pressurised containers
Nitrous oxide	Manufactured	Packaging gas	

Restrictions on Use

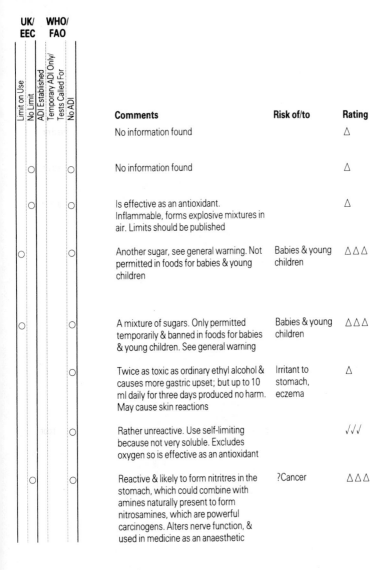

	UK/EEC		WHO/FAO			Comments	Risk of/to	Rating
	Limit on Use	No Limit	ADI Established	Temporary ADI Only/Tests Called For	No ADI			
						No information found		△
		○			○	No information found		△
		○			○	Is effective as an antioxidant. Inflammable, forms explosive mixtures in air. Limits should be published		△
	○				○	Another sugar, see general warning. Not permitted in foods for babies & young children	Babies & young children	△ △ △
	○				○	A mixture of sugars. Only permitted temporarily & banned in foods for babies & young children. See general warning	Babies & young children	△ △ △
					○	Twice as toxic as ordinary ethyl alcohol & causes more gastric upset; but up to 10 ml daily for three days produced no harm. May cause skin reactions	Irritant to stomach, eczema	△
					○	Rather unreactive. Use self-limiting because not very soluble. Excludes oxygen so is effective as an antioxidant		√√√
				○	○	Reactive & likely to form nitritres in the stomach, which could combine with amines naturally present to form nitrosamines, which are powerful carcinogens. Alters nerve function, & used in medicine as an anaesthetic	?Cancer	△ △ △

Unnumbered Additives

Names	Source	Uses	Examples
Octadecyl-ammonium acetate, Octadecylamine acetate	Manufactured	Not discovered	As an additive to the additive ammonium chloride
Oxygen	Air	Packaging gas	
Oxystearin	Synthetic	Sequestrant	Salad oils & dressings
Oxidatively polymerised soya bean oil	Soya beans	Emulsifier, stabiliser	Greasing emulsions for baking tins
Polydextrose	Manufactured	Sweetener, bulking agent	Slimming & diet foods
Propylene glycol, Propane-1,2-diol	Manufactured	Solvent, carrier for flavourings	Baked goods, sweets
Extract of quillaia	Tree bark	Surfactant, frothing agent	Soft drinks

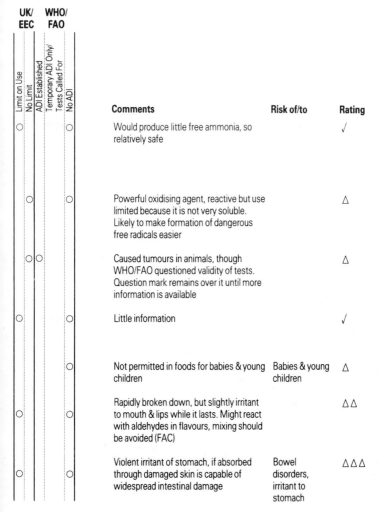

UK/EEC		WHO/FAO			Comments	Risk of/to	Rating
Limit on Use	No Limit	ADI Established	Temporary ADI Only/Tests Called For	No ADI			
O				O	Would produce little free ammonia, so relatively safe		√
	O			O	Powerful oxidising agent, reactive but use limited because it is not very soluble. Likely to make formation of dangerous free radicals easier		△
		O	O		Caused tumours in animals, though WHO/FAO questioned validity of tests. Question mark remains over it until more information is available		△
O				O	Little information		√
				O	Not permitted in foods for babies & young children	Babies & young children	△
O				O	Rapidly broken down, but slightly irritant to mouth & lips while it lasts. Might react with aldehydes in flavours, mixing should be avoided (FAC)		△ △
O				O	Violent irritant of stomach, if absorbed through damaged skin is capable of widespread intestinal damage	Bowel disorders, irritant to stomach	△ △ △

Unnumbered Additives

Names	Source	Uses	Examples
Saccharin; Sodium saccharin; Calcium saccharin	Manufactured	Intense sweetener	Wide range of slimming products, soft drinks
Spermaceti; Sperm oil	Oils from sperm whale	Releasing agent	Foods baked or moulded in tins
Aluminium potassium sulphate, Potassium aluminium sulphate, Potash alum	Manufactured mineral	Astringent	Glacé cherries
Tannic acid, Tannin	From nut galls of trees	Astringent	As flavouring in baked goods, alcoholic drinks, sweets & in other artificial flavourings
Thaumatin	Proteins from katemfe, the fruit of a Scitamineae species	Sweetener	Not widely used
Xylitol	Alcohol prepared from a simple sugar	Bulk sweetener	Not widely used

Restrictions on Use

UK/EEC			WHO/FAO		Comments	Risk of/to	Rating
Limit on Use	No Limit	ADI Established	Temporary ADI Only/Tests Called For	No ADI			
○			○		Several hundred times sweeter than sugar. 5% in the diet of rats over 2 years produces bladder tumours. Effects worse if exposed in the womb. Evidence in man is inadequate, but careful reviews support the conclusion suggested by the evidence from animals – it is a weak carcinogen on its own, & enhances the effect of other carcinogens. The ADI can be reached on only 2.2 litres of soft drinks a day – well within the range of enthusiastic consumers. Hypersensitivity & allergic reaction to light have been reported very occasionally. See also general warning	Conception, ?cancer	△ △ △
	○			○	Little information given. Vegetarians, vegans, whale lovers will wish to avoid it		√
○				○	Large doses are irritant to gums & gut, & damage muscle & kidneys if absorbed. A binge of glacé cherries might just be hazardous	Kidney disorder, irritant to stomach & mouth	△ △
	○			○	Uses at or above 1.5% can cause kidney & liver damage. Irritant to stomach; nauseating	Kidney & liver disorders, irritant to stomach	△ △
○				○	Destroyed by heat. Not permitted in foods for babies & young children	Babies & young children	△
○				○	When injected in massive doses, it is toxic to the liver, brain, urinary passages. Can cause diarrhoea when eaten. 20% is absorbed from oral doses, so safety margins would be borderline if it were ever widely used, as it may be following WHO-approved tests on it as a sugar substitute against tooth decay	Bowel disorder; babies & young children	△ △ △

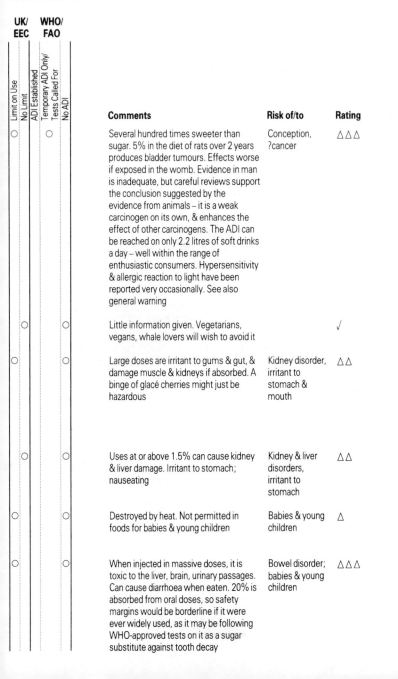

Flavourings (unnumbered)

Synthetic flavourings make up the vast number of additives in use, and they are vital to modern food processing. They are used to replace the natural flavour of food which may be lost in processing, to make up for a lack of real ingredients or for poor quality food, and as substitutes for more expensive processes such as smoking.

There are over 3,500 flavourings in use today by conservative estimates, but industry representatives put the number at over 6,000. Flavours are not subject to any regulation and do not have to be identified on labels other than by a general description.

In 1965 the Ministry of Agriculture's Food Standards Committee published a report on 'Flavouring Agents'. The Committee was told by the food industry that most flavours are used in very small quantities and that the volume and number of flavours used in food was not likely to increase much. The report noted that the trade strongly resisted the idea of a 'permitted list' of flavourings. But it said: 'We cannot accept ... that long usage and no apparent ill-effects are sufficient evidence to demonstrate that a substance is harmless in the absence of satisfactory toxicological data.' The Committee recommended that all flavours be regulated and tested. That was twelve years ago.

In fresh whole foods, flavours are a practically infallible guide to the food's quality. When we desire a food, we need it. When we have had what we need, we no longer want it. That is why technology for separating the flavour from the content of food is so dangerous for the consumer, and so highly prized by food manufacturers. It enables them to make anything palatable – but deranges the biological mechanisms which conserve our nutritional balance.

Modified Starches

Starch is useful to manufacturers as a thickening and bulking agent, but in its natural form has several technical limitations. So food technologists have devised ways to treat starch with various chemicals which improve its resistance to heat, make it more soluble in cold water and better able to produce gels, pastes or other specific textures required by industry. The starch is modified by the use of acids and alkalis and oxidising agents, which break its large molecules into smaller pieces.

Modified starches are high in calories and low in nutritional value. They are used to fill out and bind, or otherwise alter the texture of processed foods, such as soups, sauces, pie fillings and convenience meals; to stabilise whipped toppings, icings and cream fillers; to prevent ingredients separating out and a watery appearance developing, and so to lengthen shelf life; and to provide long-lasting gelling properties in a wide range of processed foods.

Modified starches need only be identified by a general description on food packets, so you have little way of telling which kind have been used. Some forms of modified starch are produced with the use of agents, epichlorohydrin and propylene oxide for example, which are suspected of causing mutations and cancer.

In 1980, a Food Additives and Contaminants Committee report on modified starches was published. It made several recommendations which have so far been ignored by government. It recommended that a 'permitted list' be drawn up, and it accepted twenty modified starches for food use. It divided these into group A substances which it held to be safe, and group B substances which needed to be tested for safety. 'We would emphasise that unless the further evidence requested is produced within three years from the publication of this report, we shall feel bound to recommend that the substances in question be removed from the permitted list.' It also recommended that 'starches modified by the use of epichlorohydrin and/or propylene oxide should not be permitted in foods described either directly or by implication as being specially prepared for infants and young children.' That was six years ago.

The starches in group B, which need more safety testing, are as follows (the E numbers are those which they will be assigned when and if the EEC produces a directive on modified starches):

E1410   Starch phosphate
E1411/2 Di-starch phosphate
E1413   Phosphated di-starch phosphate
E1430   Di-starch glycerol
E1420/1 Starch acetate
E1414   Acetylated di-starch phosphate
E1422   Acetylated di-starch adipate
E1440   Hydroxypropyl starch
E1441   Hydroxypropyl di-starch glycerol
E1442   Hydroxypropyl di-starch phosphate
E1404   Oxidised starch
–   Oxidised hydroxypropyl di-starch glycerol
–   Sodium octenyl succinated starch

Modified starches are not only potential health hazards, they also deceive the consumer. Michael Jacobson, an American expert on additives quotes a clear example: in 1985 a well-known baby food company in the USA announced that it was removing modified starches from its baby foods, despite the fact that other manufacturers were saying it was simply not possible to make baby foods without them. To promote its newly formulated products, the company ran an advertising campaign explaining that, because the starches which previously added empty bulk had been removed, its baby foods would now contain 25 per cent more vegetables and 50 per cent more meat.

Enzymes

Enzymes are used in the manufacture of a wide range of foods, but they are nearly always processing aids, and so do not appear on labels.

They are used in brewing and baking, cheese-making, in refining sugar and in producing modified starches. They are also used to tenderise meat – twenty-four hours before slaughter cattle are injected with an enzyme which digests the animals' tissues to give ready-tenderised meat. Another function is to clean up contaminated milk. When cows have been injected with antibiotics, residues may be found in the milk, and farmers are meant to discard it as unfit for human consumption. But they sometimes add the enzyme penicillinase to the milk instead.

A government report on enzymes, published in 1982 by the Food Additives and Contaminants Committee said, 'We do not agree that the addition of one substance to milk to remove another is preferable to discarding the contaminated milk.'

The report made several recommendations which have not been implemented, including the banning of penicillinase. Most enzymes used today are produced by microbial fermentation, a few are derived from animal or plant tissue. The report says that while only small residues of enzymes remain in food, those produced microbially can contain toxins, and further safety tests are required. 'Although we have been assured by industry that only non-toxic-producing strains are used for enzyme production, we have received no satisfactory documentary evidence to this effect. Neither do we accept the claim that natural occurrence of the source organism in food is necessarily sufficient evidence of safety.'

The report also covers allergies to enzymes: 'We have found no evidence of allergenicity among consumers of food processed with enzymes . . . we do not find this surprising, however, as we are not aware of any facilities which would enable such cases to be identified.' There are, however, plenty of reports of allergy among workers exposed to enzymes.

The conclusion of the report is that all enzymes should be regulated by a 'permitted list', and that many of the microbially produced enzymes should be put in group B because not enough testing has been done to establish their safety, and that their use should be reviewed within two years. That was four years ago.

Alphabetical List of Additives

△ **E414**	Acacia gum
△△△ **–**	Acesulfame potassium
√√ **E260**	Acetic acid
△ **E472(a)**	Acetic acid esters of mono and diglycerides of fatty acids
△ **E472(a)**	Acetoglycerides
△ **E472(e)**	mono and diAcetyltartaric acid esters of mono and diglycerides of fatty acids
△△ **E142**	Acid brilliant green
√√ **E341(a)**	Acid calcium phosphate
√√ **E339(b)**	Acid sodium phosphate
△△△ **E450(a)**	Acid sodium pyrophosphate
△△△ **E222**	Acid sodium sulphite
√√ **E341(a)**	ACP
√√ **355**	Adipic acid
√√ **E406**	Agar, agar-agar
△ **E405**	Alginate ester
√√√ **E400**	Alginic acid
√√ **E160(a)**	Alpha-carotene
△ **E460(ii)**	Alpha-cellulose
√√√ **E307**	Alpha-tocopherol
△△ **E173**	Aluminium
△ **556**	Aluminium calcium silicate
△△ **–**	Aluminium potassium sulphate
△ **554**	Aluminium sodium silicate
△△△ **E123**	Amaranth
√√ **E440(b)**	Amidated pectin
△ **–**	2-Aminoethanol
△ **E403**	Ammonium alginate
△ **503**	Ammonium bicarbonate
△ **503**	Ammonium carbonate
△ **510**	Ammonium chloride
△ **380**	triAmmonium citrate
△ **–**	Ammonium dihydrogen orthophosphate (see E341)
△ **–**	diAmmonium hydrogen orthophosphate (see E341)
△ **381**	Ammonium ferric citrate
△ **381**	Ammonium ferric citrate, green
△ **503**	Ammonium hydrogen carbonate
△△ **527**	Ammonium hydroxide
√√ **E440(a)**	Ammonium pectate
△ **–**	Ammonium phosphate dibasic (see E341)
△ **–**	Ammonium phosphate monobasic (see E341)
√ **442**	Ammonium phosphates
△ **545**	Ammonium polyphosphates
△ **–**	Ammonium sulphate (see 516)
△ **E160(b)**	Annatto
√√ **E163**	Anthocyanins
√√√ **E300**	L-Ascorbic acid
√√ **E304**	Ascorbyl palmitate
△△△ **–**	Aspartame
√ **927**	Azodicarbonamide
√ **927**	Azoformamide
△△△ **E122**	Azorubine
√ **500**	Baking soda
√√√ **E162**	Beetroot red
√ **901**	Beeswax, yellow
√ **901**	Beeswax, white
√ **558**	Bentonite, bentonitum
△△ **E210**	Benzoic acid
√√ **E160(e)**	Beta-8-apocarotenal
√√ **E160(e)**	Beta-apo-8'-carotenal
√√ **E160(f)**	Beta-apo-8'-carotenal, ethyl ester of
√√ **E160(a)**	Beta-carotene
√√√ **E162**	Betanin
△△△ **E320**	BHA
△△△ **E321**	BHT
√ **500**	Bicarbonate of soda
△△△ **E230**	Biphenyl
△ **E161(b)**	Bixin
△△ **E151**	Black PN
△△ **E151**	Brilliant black BN
△△△ **133**	Brilliant blue FCF
△△△ **154**	Brown FK
△△△ **155**	Brown HT
△△△ **E320**	Butylated hydroxyanisole
△△△ **E321**	Butylated hydroxytoluene
√ **–**	Butyl stearate (see 570)
√ **E263**	Calcium acetate
√ **E404**	Calcium alginate
△ **556**	Calcium aluminium silicate
△△ **E213**	Calcium benzoate
△△△ **E227**	Calcium bisulphite
√√ **E170**	Calcium carbonate
△ **509**	Calcium chloride
√√ **E333**	Calcium citrate
√√ **E333**	monoCalcium citrate
√√ **E333**	diCalcium citrate

√√ **E333**	triCalcium citrate	
√ **623**	Calcium dihydrogen di-L-glutamate	
√√ **E341(c)**	triCalcium diorthophosphate	
√√√ **540**	diCalcium diphosphate	
△△ **385**	Calcium disodium EDTA	
△△ **385**	Calcium disodium ethylene-diamine-NNN'N'-tetra-acetate	
△△△ **E238**	Calcium formate	
√ **578**	Calcium gluconate	
√ **623**	Calcium glutamate	
△ −	Calcium heptonate	
√√ **352**	Calcium hydrogen malate	
√√ **E341(b)**	Calcium hydrogen orthophosphate	
√√√ **540**	Calcium hydrogen phosphate	
△△△ **E227**	Calcium hydrogen sulphite	
√√ **526**	Calcium hydroxide	
△ **E327**	Calcium lactate	
√√ **E302**	Calcium L-ascorbate	
√√ **352**	Calcium malate	
√ −	Calcium mesoinositol-hexaphosphate	
√ **529**	Calcium oxide	
√√ **E341(c)**	triCalcium phosphate	
√√ **E341(b)**	Calcium phosphate dibasic	
√ −	Calcium phytate	
△ **544**	Calcium polyphosphates	
△ **E282**	Calcium propionate	
△ **E470**	Calcium salts of fatty acids	
△△△ −	Calcium saccharin	
√ **552**	Calcium silicate	
△△ **E203**	Calcium sorbate	
△ −	Calcium stearate (see 572)	
√ **E482**	Calcium stearoyl-2-lactylate	
√ **516**	Calcium sulphate	
△△△ **E226**	Calcium sulphite	
√√ **E341(a)**	Calcium tetrahydrogen diorthophosphate	
△ **E161(g)**	Canthaxanthin	
√√ **E160(c)**	Capsanthin	
√√ **E160(c)**	Capsorubin	
△△△ **E150**	Caramel	
△△△ **E153**	Carbon black	
√ **E290**	Carbon dioxide	
△ **E466**	Carboxymethylcellulose, sodium salt of	
△ **E466**	Carmellose sodium	
△△ **E120**	Carminic acid	

△△△ **E122**	Carmoisine	
√√√ **903**	Carnauba wax	
√√√ **E410**	Carob gum	
√√ **E160(e)**	beta-apo-8'-Carotenal	
√√ **E160(a)**	Carotene (alpha, beta and gamma)	
△△△ **E407**	Carrageenan	
△ **E460(ii)**	Cellulose, alpha	
△ **E466**	Cellulose, carboxymethyl	
△ **E465**	Cellulose, ethylmethyl	
△ **E463**	Cellulose, hydroxypropyl	
△ **E464**	Cellulose, hydroxypropylmethyl	
△ **E461**	Cellulose, methyl	
△ **E465**	Cellulose, methylethyl	
△ **E460(i)**	Cellulose, microcrystalline	
△ **E460(ii)**	Cellulose, powdered	
√√ **E170**	Chalk	
△△△ **E251**	Chile saltpetre	
△△△ **925**	Chlorine	
△△△ **926**	Chlorine dioxide	
△ −	diChlorodifluoromethane	
△△△ −	diChloromethane	
√√ **E140**	Chlorophyll	
√√ **E141**	Chlorophyllins	
△△△ **155**	Chocolate brown HT	
√√ **E330**	Citric acid	
△ **E472(c)**	Citric acid esters of mono and diglycerides of fatty acids	
△ **E472(c)**	Citroglycerides	
△ **E412**	Cluster bean	
△ **E466**	CMC	
△△ **E120**	Cochineal	
△△△ **E124**	Cochineal Red R	
△ **E461**	Cologel	
√√ **E415**	Corn sugar gum	
√√ **E336**	Cream of tartar	
√ **E161(c)**	Cryptoxanthin	
△△△ **E100**	Curcumin	
√√ **E163(a)**	Cyanidin	
√√√ **920**	L-Cysteine hydrochloride	
√√√ **920**	L-Cysteine hydrochloride monohydrate	
√√ **E163(b)**	Delphinidin	
√√√ **E309**	Delta tocopherol	
△△△ −	Dextrose	
√ **575**	D-glucono-1,5-lactone	
△△ −	Diacetin	
△ **E472(e)**	Diacetyltartaric acid esters of mono and diglycerides of fatty acids	

△ –	Diammonium hydrogen orthophosphate (see E341)	
√√ **E333**	Dicalcium citrate	
△ –	Dichlorodifluoromethane	
△△△ –	Dichloromethane	
△ –	Diethyl ether	
△ **E470**	Diglycerides of fatty acids	
△△△ **900**	Dimethicone	
△△△ **900**	Dimethylpolysiloxane	
△△△ –	Dioctyl sodium sulphosuccinate	
△△△ **E230**	Diphenyl	
√√ **E331(b)**	Disodium citrate	
△△△ **E450(a)**	Disodium dihydrogen diphosphate	
△△ –	Disodium dihydrogen ethylene-diamene-NNN'N'-tetra-acetate (see 385)	
√√√ –	Disodium DL-tartrate (see E335)	
△△ –	Disodium edetate (see 385)	
△ **627**	Disodium guanylate	
√√ **E339(b)**	Disodium hydrogen phosphate	
√√√ **E335**	Disodium L-(+)-tartrate	
△△△ **E223**	Disodium pyrosulphite	
√√ –	Dipotassium DL-tartrate (see E336)	
√√ **E340(b)**	Dipotassium hydrogen orthophosphate	
√√ **E336**	Dipotassium tartrate	
△ **296**	DL-malic acid	
√√ **E334**	DL-tartaric acid	
△△△ **E312**	Dodecyl gallate	
△△△ **E312**	Dodecyl 3,4,5 trihydroxybenzene	
√ **542**	Edible bone phosphate	
√ **442**	Emulsifier YN	
△△△ **E127**	Erythrosine BS	
√ –	Ethanol	
△ –	monoEthanolamine	
△ –	Ethoxyquin	
△ –	Ethyl acetate	
√ –	Ethyl alcohol	
√√ **E160(f)**	Ethyl ester of beta-apo-8'-carotenal (E160(e))	
△ –	diEthyl ether	
△△ **E214**	Ethyl 4-hydroxybenzoate	
△△ **E215**	Ethyl 4-hydroxybenzoate, sodium salt of	
△ **637**	Ethyl maltol	
△ **E465**	Ethylmethylcellulose	
△△ **E214**	Ethyl parahydroxybenzoate	
△△△ –	Extract of quillaia	
△ **381**	Ferric ammonium citrate	
√ **E161(a)**	Flavoxanthin	
△△△ **154**	Food brown	
△△△ **E236**	Formic acid	
△△△ –	Fructose	
△ **297**	Fumaric acid	
△△△ **E310–2**	Gallates	
√√ **E160(a)**	Gamma-carotene	
√√√ **E308**	Gamma-tocopherol	
√ **575**	Glucono deltalactone	
√ **575**	D-Glucono-1,5-lactone	
△△△ –	Glucose syrup, dried/hydrogenated	
√ **620**	L-Glutamic acid	
√ **E471**	mono and diGlycerides of fatty acids	
△ **E422**	Glycerol	
△ **E472(a)**	Glycerol esters	
–	polyGlycerol esters of dimerised fatty acids of soya bean oil (see 476)	
△ **E475**	polyGlycerol esters of fatty acids	
△△ **E476**	polyGlycerol esters of polycondensed fatty acids of castor oil	
△△ –	Glycerol monoacetate	
△△ –	Glycerol diacetate	
△△ –	Glycerol triacetate	
√ **E471**	Glyceryl distearate	
√ **E471**	Glyceryl monostearate	
√√ –	Glycine	
√ **E175**	Gold	
△△ **E142**	Green S	
△ **627**	Guanosine 5'-disodium phosphate	
△ **E412**	Guar flour, guar gum	
△ **E414**	Gum acacia	
△ **E414**	Gum arabic	
△ **E413**	Gum dragon	
△ **E414**	Gum hashab	
△ **E413**	Gum tragacanth	
√ **516**	Gypsum	
√ **503**	Hartshorn	
△ **370**	1,4-Heptonolactone	
△△△ **E239**	Hexamine	
△△△ **E239**	Hexamethylene tetramine	
√√ **355**	Hexanedioic acid	
△ **507**	Hydrochloric acid	

√√ **375** Nicotinamide
√√ **375** Nicotinic acid
√ **234** Nisin
√√√ **–** Nitrogen
△△△ **–** Nitrous oxide
△ **E160(b)** Norbixin
√ **–** Octadecylammonium acetate
√ **–** Octadecylamine acetate
△△△ **E311** Octyl gallate
√√ **E304** 6-O-palmitoyl L-ascorbic acid
△△△ **E231** O-phenyl phenol
△△△ **E110** Orange yellow S
△△△ **E231** Orthophenyl phenol
√√ **E338** Orthophosphoric acid
√ **–** Oxidatively polymerised soya bean oil
△△△ **432** polyOxyethylene sorbitan monolaurate
△△△ **433** polyOxyethylene sorbitan mono-oleate
△△△ **434** polyOxyethylene sorbitan monopalmitate
△△△ **435** polyOxyethylene sorbitan monostearate
△△△ **436** polyOxyethylene sorbitan tristearate
△△△ **430** polyOxyethylene (8) stearate
△△△ **431** polyOxyethylene (40) stearate
△ **–** Oxygen
△△△ **430** polyOxyl (8) stearate
△△△ **431** polyOxyl (40) stearate
△ **–** Oxystearin
√√ **E304** 6-O-Palmitoyl L-ascorbic acid
△△△ **E131** Patent blue V
√√ **E440(a)** Pectin
√√ **E440(b)** Pectin, amidated
√√ **E440(b)** Pectin extract
√√ **E163(d)** Pelargonidin
△△△ **E450(b)** Pentapotassium triphosphate
△△△ **E450(b)** Pentasodium triphosphate
√√ **E163(e)** Peonidin
√ **530** Periclase
√√ **E163(f)** Petunidin
△△△ **E230** biPhenyl
△△△ **E231** O-Phenyl phenol
△△△ **E231** orthoPhenyl phenol

√√ **E338** Phosphoric acid
△△ **E180** Pigment rubine
√ **516** Plaster of Paris
△ **–** Polydextrose
△ **E475** Polyglycerol esters of fatty acids
△△ **476** Polyglycerol esters of polycondensed fatty acids of castor oil
△△ **476** Polyglycerol polyricinoleate
△△△ **432** Polyoxyethylene sorbitan monolaurate
△△△ **433** Polyoxyethylene sorbitan mono-oleate
△△△ **434** Polyoxyethylene sorbitan monopalmitate
△△△ **435** Polyoxyethylene sorbitan monostearate
△△△ **436** Polyoxyethylene sorbitan tristearate
△△△ **430** Polyoxyethylene (8) stearate
△△△ **431** Polyoxyethylene (40) stearate
△△△ **430** Polyoxyl (8) stearate
△△△ **431** Polyoxyl (40) stearate
△ **545** Polyphosphates, ammonium
△ **544** Polyphosphates, calcium
△△△ **E450(c)** Polyphosphates, potassium
△△△ **E450(c)** Polyphosphates, sodium
△△△ **432** Polysorbate 20
△△△ **434** Polysorbate 40
△△△ **435** Polysorbate 60
△△△ **436** Polysorbate 65
△△△ **433** Polysorbate 80
△△△ **E124** Ponceau 4R
△△ **–** Potash alum
√√ **E261** Potassium acetate
√√ **E336** Potassium acid L-(+)-tartrate
√ **E402** Potassium alginate
△△ **–** Potassium aluminium sulphate
△△ **E212** Potassium benzoate
△△△ **924** Potassium bromate
√ **501** Potassium carbonate
△ **508** Potassium chloride
√√ **E332** Potassium citrate
√√ **E332** monoPotassium citrate
√√ **E332** triPotassium citrate
√√ **E332** Potassium dihydrogen citrate

√√ **E340(a)**	Potassium dihydrogen orthophosphate	
△△△ **E450(a)**	tetraPotassium diphosphate	
√√√ –	Potassium disodium DL-tartrate (see E337)	
√√ –	diPotassium DL-tartrate (see E336)	
√√ –	monoPotassium DL-tartrate (see E336)	
△△△ **536**	Potassium ferrocyanide	
√ **577**	Potassium gluconate	
√ **622**	monoPotassium glutamate	
√ **536**	Potassium hexacyanoferrate II	
√ **501**	Potassium hydrogen carbonate	
√ **622**	Potassium hydrogen L-glutamate	
√√ **E336**	Potassium hydrogen L-(+)-tartrate	
√√ **E340(b)**	diPotassium hydrogen orthophosphate	
√√ **E336**	Potassium hydrogen tartrate	
△△ **525**	Potassium hydroxide	
△ **E326**	Potassium lactate	
√√ **E336**	monoPotassium L-(+)-tartrate	
√√ **351**	Potassium malate	
△△△ **E224**	Potassium metabisulphite	
△△△ **E252**	Potassium nitrate	
△△△ **E249**	Potassium nitrite	
√√ **E340(c)**	triPotassium orthophosphate	
√√ **E440(a)**	Potassium pectate	
√√ **E340(b)**	Potassium phosphate dibasic	
√√ **E340(a)**	Potassium phosphate monobasic	
√√ **E340(c)**	Potassium phosphate tribasic	
△△△ **E450(c)**	Potassium polyphosphates	
√ **E283**	Potassium propionate	
△△△ **E224**	Potassium pyrosulphite	
△ **E470**	Potassium salts of fatty acids	
√√√ **E337**	Potassium sodium L-(+)-tartrate	
△△ **E202**	Potassium sorbate	
√ **515**	Potassium sulphate	
√√ **E336**	diPotassium tartrate	
△△△ **E450(b)**	pentaPotassium triphosphate	
△△△ **E450(b)**	Potassium tripolyphosphate	
△ **E460(ii)**	Powdered cellulose	
△△ –	Propane -1,2-diol	
△ **E405**	Propane -1,2-diol alginate	
△ **E477**	Propane -1,2-diol esters of fatty acids	
√ **E280**	Propionic acid	
△△ –	Propylene glycol	
△ **E405**	Propylene glycol alginate	
△ **E477**	Propylene glycol esters of fatty acids	
△△△ **E310**	Propyl gallate	
△△ **E216**	Propyl 4-hydroxybenzoate	
△△ **E217**	Propyl 4-hydroxybenzoate, sodium salt of	
△△ **E216**	Propyl parahydroxybenzoate	
△△△ **E310**	Propyl 3,4,5 trihydroxybenzene	
△△△ –	Quillaia, extract of	
△△ **E104**	Quinoline yellow	
△△△ **128**	Red 2G	
△△△ **907**	Refined microcrystalline wax	
√ **E161(f)**	Rhodoxanthin	
√√ **E101**	Riboflavin	
√√ **101(a)**	Riboflavin-5′-phosphate	
√√√ **E337**	Rochelle salt	
△△ **E180**	Rubine	
√ **E161(d)**	Rubixanthin	
△△△ –	Saccharin	
△ **E470**	Salts of fatty acids	
√ **904**	Shellac	
√ **551**	Silica, silicea	
√ **551**	Silicon dioxide	
△△ **E174**	Silver	
△△△ **900**	Simethicone	
△ **558**	Soap clay	
△ **E470**	Soaps	
√ **262**	Sodium acetate	
√√√ **E401**	Sodium alginate	
△ **541**	Sodium aluminium phosphate, acidic/basic	
△△ **E211**	Sodium benzoate	
√ **500**	Sodium bicarbonate	
△△△ **E232**	Sodium biphenyl-2-yl oxide	
△△△ **E222**	Sodium bisulphite	
√ **500**	Sodium carbonate	
△ **E466**	Sodium carboxymethyl cellulose	
√√ **E331**	Sodium citrates	

√√	**E331(b)**	diSodium citrate
√√	**E331(c)**	triSodium citrate
√√	**E262**	Sodium diacetate
√	**E331(a)**	Sodium dihydrogen citrate
△△△	**E450(a)**	diSodium dihydrogen diphosphate
√√√	–	diSodium dihydrogen ethylene diamene – NNN'N'-tetra-acetate (see 385)
√√	**E339(a)**	Sodium dihydrogen orthophosphate
△△△	**E450(a)**	tetraSodium diphosphate
△△△	**E450(a)**	triSodium diphosphate
√√√	–	diSodium D L-tartrate (see E335)
√√√	–	monoSodium D L-tartrate (see E335)
√√√	–	diSodium edetate (see 385)
△△	**E215**	Sodium ethyl parahydroxybenzoate
△△△	**535**	Sodium ferrocyanide
△△△	**E237**	Sodium formate
√√√	**576**	Sodium gluconate
△△△	**621**	monoSodium glutamate (MSG)
△	**627**	Sodium guanylate
△	**627**	diSodium guanylate
△	–	Sodium heptonate
△△△	**535**	Sodium hexacyanoferrate II
√√	**E262**	Sodium hydrogen acetate
√	**500**	Sodium hydrogen carbonate
√√	**E262**	Sodium hydrogen diacetate
△△△	**621**	Sodium hydrogen L-glutamate
√√	**350**	Sodium hydrogen malate
√√	**E339(a)**	Sodium hydrogen orthophosphate
√√	**E339(b)**	diSodium hydrogen phosphate
△△△	**E222**	Sodium hydrogen sulphite
△△	**524**	Sodium hydroxide
△	**631**	Sodium 5'-inosinate
△	**E325**	Sodium lactate
√√√	**E301**	Sodium L-ascorbate
√√√	**E335**	diSodium L-(+)-tartrate
√√√	**E335**	monoSodium L-(+)-tartrate
√√	**350**	Sodium malate
△△△	**E223**	Sodium metabisulphite
△△	**E219**	Sodium methyl para-hydroxybenzoate
△△△	**E251**	Sodium nitrate
△△△	**E250**	Sodium nitrite
△△△	**E232**	Sodium orthophenylphenate
△△△	**E232**	Sodium orthophenylphenol
-√√	**E339(c)**	triSodium orthophosphate
√√	**E440(a)**	Sodium pectate
△△△	**E450(c)**	Sodium polyphosphates
√√√	**E337**	Sodium potassium L-(+)-tartrate
△	**E281**	Sodium propionate
△△	**E217**	Sodium propyl parahydroxybenzoate
△△△	**E450(a)**	tetraSodium pyrophosphate
△△△	**E223**	diSodium pyrosulphite
△	**635**	Sodium 5'-ribonucleotide
△△△	–	Sodium saccharin
△	**E466**	Sodium salt of carboxymethyl cellulose
△△	**E215**	Sodium salt of ethyl 4-hydroxybenzoate (E214)
△△	**E219**	Sodium salt of methyl 4-hydroxybenzoate (E218)
△△	**E217**	Sodium salt of propyl 4-hydroxybenzoate (E216)
△	**E470**	Sodium salts of fatty acids
√	**500**	Sodium sesquicarbonate
△△	**E201**	Sodium sorbate
√	**E481**	Sodium stearoyl-2-lactylate
△	**514**	Sodium sulphate
△△	**E221**	Sodium sulphite
△△△	**E450(b)**	pentaSodium triphosphate
△△△	**E450(b)**	Sodium tripolyphosphate
△	–	Solvent ether
△△	**E200**	Sorbic acid
△△△	**493**	Sorbitan monolaurate
△△	**494**	Sorbitan mono-oleate
△△	**495**	Sorbitan monopalmitate
△△	**491**	Sorbitan monostearate
△△	**492**	Sorbitan tristearate
△△	**E420(i)**	Sorbitol
△△	**E420(ii)**	Sorbitol syrup
△△	**493**	Span 20
△△	**495**	Span 40
△△	**492**	Span 65
△△	**494**	Span 80
√	–	Spermaceti
√	–	Sperm oil
√√√	**570**	Stearic acid
√	**E483**	Stearyl tartrate
√	**E416**	Sterculia gum
√	**363**	Succinic acid
△	**E474**	Sucroglycerides

△△△ –	Sucrose	
△ **E473**	Sucrose esters of fatty acids	
△ **E414**	Sudan gum	
△△△ –	Sugar	
△△△ **E220**	Sulphur dioxide	
△△△ **513**	Sulphuric acid	
△△△ **E110**	Sunset yellow FCF	
△△△ **553(b)**	Talc	
△△ –	Tannic acid, tannin	
√√ **E334**	DL-Tartaric acid	
√√ **E334**	L-(+)-Tartaric acid	
△ **E472(d)**	Tartaric acid esters of mono and diglycerides of fatty acids	
△△ **E102**	Tartrazine	
√√ **E341(c)**	TCP	
△△△ **E450(a)**	Tetrasodium diphosphate	
△△△ **E450(a)**	Tetrasodium pyrophosphate	
△ –	Thaumatin	
△ **E233**	Thaibendazole	
△ **E233**	2(Thiazol-4-yl) benzimidazole	
√ **E171**	Titanium dioxide	
√√√ **E307**	alpha-Tocopherol	
√√√ **E308**	gamma-Tocopherol	
√√√ **E309**	delta-Tocopherol	
√√√ **E306**	Tocopherols	
△ **E413**	Tragacanth	
△△ –	Triacetin	
△ **E380**	Triammonium citrate	

√√ **E333**	Tricalcium citrate	
√√ **E341(c)**	Tricalcium diorthophosphate	
√√ **E341(c)**	Tricalcium phosphate	
√√ **E332**	Tripotassium citrate	
√√ **E340(c)**	Tripotassium orthophosphate	
√√ **E331(c)**	Trisodium citrate	
△△△ **E450(a)**	Trisodium diphosphate	
√√ **E339(c)**	Trisodium orthophosphate	
√ **500**	Trona	
△ **E100**	Turmeric	
△△△ **432**	Tween 20	
△△△ **434**	Tween 40	
△△△ **435**	Tween 60	
△△△ **436**	Tween 65	
△△△ **433**	Tween 80	
△△△ **E153**	Vegetable carbon	
√ **E161(e)**	Violoxanthin	
√√ **375**	Vitamin B	
√√ **E101**	Vitamin B_2	
√√√ **E300**	Vitamin C	
√√√ **E306**	Vitamin E	
△△△ **907**	Wax, refined microcrystalline	
√√ **E415**	Xanthan gum	
√ **E161**	Xanthophylls	
△△△ –	Xylitol	
△△△ **107**	Yellow 2G	

How Many Additives Do You Eat?

Research by Adriana Luba

We bought a shopping basket of foods which might make up a typical day's eating, and counted the number of additives it contained: grand total – 150. And those are just the additives which are declared on the labels. There's no way of counting all the additives used as processing aids which don't have to be declared, or the residues of chemicals from farming or in the water in your tea. You may be getting more than you bargained for!

Breakfast
Instant orange juice
Crunchy oat cereal with milk
2 slices of white bread toast
Low fat spread
Jam (two types)
Coffee
Coffee whitener
Sugar

Elevenses
Cup of tea
Chocolate wafer biscuit
Chocolate cake biscuits

Sandwich Lunch   or   *Home Lunch*
Two rolls   Instant snack meal
spread with margarine,   Cheese on toast
Ham & tongue spreading pate,   Strawberry dessert
Salmon paste   Apple juice
Fruit pie
Soft drink in a can
Cheesy corn snacks

Afternoon Break
Tea
Slice of Swiss roll

Supper
Packet mushroom soup
Turkey steaklettes
Instant mashed potato
Peas
Breaded cauliflower
Gravy
Trifle
Orange drink

Bedtime Drink and Snack
Hot chocolate
Slice of malt loaf

Details of Meals and Additives

The details given below are taken in each case from the labels of well known branded products. The number of different additives and the number of 'doses' (i.e. total number of times additives are declared) are given for each meal, with snacks listed separately, and for the day as a whole.

Breakfast

Instant orange juice	Colours	E102, E110
	Antioxidant	E320
	Stabiliser	E466
Crunchy oat cereal	Colours	E150, E160(b)
	Raising agents	E450(a), 500
White sliced bread	Preservatives	E280, E281
	Emulsifier	E471
	Flour improvers	E300, 920, 924, 926, 927
Low fat spread	Colour	E160(a)
	Preservative	E202
	Antioxidant synergists	E325, E331
	Emulsifier	E339, E471

Strawberry jam	Colour	E124
	Acidity regulator	296
	Antioxidant synergist	E341
	Preservative	E202
	Gelling agent	E440(b)
Blackcurrant jam	Colours	E122, E132
	Acidity regulator	296
	Antioxidant synergist	E341
	Preservative	E202
	Gelling agent	E440(b)
Coffee whitener	Colours	E102, E110
	Emulsifiers	E340, E471, E472(e)

Total number of doses in breakfast **38**
Number of different additives **29**

Sandwich Lunch

White bread rolls	Preservative	E282
	Emulsifiers	E471, E472(e)
	Flour improvers	E300, 920, 924, 926, 927
Soft margarine	Colour	E160(a)
	Emulsifiers	E322, E471
Ham & tongue spreading paté	Colours	E102, E128
	Preservative	E250
	Emulsifier	E450(c)
	Flavour enhancer	621
Salmon paste	Colour	E172
	Antioxidant	E300
Apple & blackcurrant pie	Colours	E122, E142
	Preservatives	E202, E330
	Emulsifier	E471
Pineapple & grapefruit drink in a can	Colours	E102, E110
	Preservatives	E211, E330
Cheesy corn snacks	Colours	E102, E110
	Flavour enhancer	621

Total number of doses in sandwich lunch **30**
Number of different additives **22**

Lunch At Home

Instant snack meal	Colours	E102, E123, E124, E150, 154
	Preservatives	E220, E262
	Antioxidants	E320, E321, E331
	Flavour enhancers	621, 627, 631
White bread	Preservatives	E280, E281
	Emulsifier	E471
	Flour improvers	E300, 920, 924, 926, 927
Soft margarine	Colour	E160(a)
	Emulsifiers	E322, E471
Processed cheddar	Preservative	E200
	Antioxidant synergist	E331
	Emulsifiers	E339, E401
Strawberry dessert	Colour	E124
	Preservative	E202
	Antioxidant synergist	E330
	Emulsifiers	E401, E407
Apple juice drink	Colours	E102, E123, E142
	Acid	296
	Antioxidant	E300

Total number of doses in lunch at home	38
Number of different additives	31

Supper

Packet cream of mushroom soup	Antioxidants	E320, E321
	Emulsifiers	E340, E450(c), E471, E472
	Flavour enhancer	635
Turkey steaklettes	Emulsifiers	E322, E450(b)
	Anti-caking agent	552
	Flavour enhancer	621
Instant mashed potato	Preservative	E223
	Antioxidants	E320, E321
	Emulsifiers	E450(a), E471

Tinned peas	Colours	E102, E142
Breaded cauliflower	Emulsifier	541
	Raising agents	E450(a), 500
Gravy granules	Colour	E150
	Antioxidants	E320, E321
	Emulsifiers	E472(e), E482
	Anti-caking agent	554
	Flavour enhancers	621, 627, 631
Table salt	Anti-caking agent	504, 535
Trifle	Colours	E102, E110, E122, E123, E124, E127, E132, E160(a)
	Preservative	E202
	Antioxidants	E320, E340
	Acidity regulator	355
	Gelling agents	E410, 508
	Thickener	E466
	Emulsifiers	E477, E407
Whole orange drink	Colours	E104, E110
	Preservatives	E211, E223
	Stabiliser	E466

Total number of doses in supper	**51**
Number of different additives	**40**

Snacks

Elevenses

Chocolate coated wafer biscuit	Colour	E150
	Antioxidant	E320
	Emulsifier	E322
Chocolate cake biscuits	Colours	E102, E110
	Acidity regulators	E330, E331
	Gelling agent	E440
	Raising agents	E450(a), 500, 503

Afternoon Break

Chocolate Swiss roll	Colour	E150
	Emulsifiers	E322, E471

Bedtime Drink and Snack

Hot chocolate drink	Colours	E102, E110
	Emulsifiers	E340, E466, E471
		E472(e)
Fruit malt loaf	Colour	E150
	Preservatives	E202, E282

Total number of doses in snacks	23
Number of different additives	17
Total number of doses in complete day's eating	151
Number of different additives	62

The above figures apply to those who decided to have lunch at home. The number of doses (151) includes 39 colourings, led by tartrazine (E102), sunset yellow FCF (E110) and caramel (E150); 19 preservatives, led by potassium sorbate (E202); and 14 antioxidants, led by BHA (E320) and BHT (E321) – some of the most commonly used additives are also among the most suspect.

Taking the sandwich lunch instead, the number of doses comes out at 136 and the number of different additives at 51, but the pattern is much the same.

Guide to Safer Shopping

Research by Adriana Luba

By cutting down on harmful additives wherever possible, you can reduce the toxic load you face. Not all additives have to be declared on food labels – anything which is a processing aid need not appear – e.g. bleach used in flour which is then used for making biscuits would not appear on the biscuit label, nor would releasing agents used to stop the product sticking to machinery during processing. It is, therefore, not possible to guarantee that the products listed here are entirely free of suspect additives. Nevertheless, by buying them in preference to other additive-laden brands, you can reduce considerably the number of chemicals you eat in your food.

Manufacturers may change the ingredients of their products – so it is always worth checking the labels and watching out for any new additions to the list of foods which have had harmful additives removed. The ingredients of all the products listed here have been checked by us and were correct at the time of going to press.

We have listed products which are free from those additives which we have rated as suspect in our numerical guide. Nearly all retailers also stock several types of food which are generally additive-free, such as fresh fruit and vegetables, fresh meat, dried beans and pulses, grains and rice, nuts, pasta and fresh dairy produce, and these are not included here.

In some cases we have listed products with additives which have question marks over them, modified starches and flavourings, for example (marked with a $\triangle$). We have included them where they present the best choice among a limited range – as in convenience meals – and we have listed the additives they contain. This list is not comprehensive – new additive-free foods are appearing on the market all the time. We would be delighted to hear from consumers, retailers and manufacturers about any other products which deserve to be included in this list, and would like to thank all those who have supplied information.

Breakfast Cereals
There is a wide range of breakfast cereals available in all the large

supermarket chains which do not contain additives other than vitamins and minerals. Choose varieties which are low in added sugars and salt. Also watch out for mueslis which have sulphur dioxide (E220), added as a preservative to the fruit. Own-label porridge oats are almost without exception additive-free, and so have not been listed separately below.

Boots
Muesli
Honey Muesli
No Sugar Added Muesli

British Home Stores
Bran Crunch with Apple
Honey Crunch
Swiss Style Muesli

Co-Op
Cornflakes
Swiss Style Cereal
Wholewheat Biscuits
Wholewheat Flakes

Fine Fare
Crunchy Breakfast Cereal
Morning Bran
Muesli
Wholewheat Breakfast Biscuits
Wholefood Muesli
Wheatflakes

Presto
Cornflakes
Rice Crunchies
Swiss Style Breakfast Cereal
Wholewheat Breakfast Bisk

Safeway
Cornflakes
Crunchy Breakfast Cereal
Fibre Bran
Hot Oat Cereal   Modified starch
Quick Cooking Oats

Rice Crunchies
Swiss Style Cereal
Wholewheat Flakes & Biscuits

Sainsbury
Bran Flakes
Bran Muesli
Cornflakes
Crunchy Oat Cereal
Crunchy Oat, Apple & Bran
Muesli
Wholewheat Bisk
Wheatflakes

Tesco
Branflakes
Coco Puffs
Cornflakes
Crunchy Nut & Honey Cereal
Puffed Rice
Swiss Style Cereal
Wholewheat Cereal

Waitrose
Branflakes
Cornflakes
Wholewheat Biscuits

Jordans
Country Muesli
Original Crunchy Toasted Oat Cereal: Natural
Bran & Apple
Honey Almond & Raisins

Special Recipe Muesli

Kellogg's
All-Bran (but *not* Bran Buds which contain caramel)
Bran Flakes
Coco Pops
Cornflakes
Country Store
Crunchy Nut Cornflakes
Frosties

Fruit 'n' Fibre
Ricicles
Rice Krispies
Special K
Start
Sultana Bran
Summer Orchard

Nabisco
Cubs
Harvest Home Bran Flakes
Shredded Wheat
Shreddies

Weetabix
Alpen
Bran Fare
Weetabix
Wheetaflakes
Wheetaflakes 'n' Raisins

The following brand names are available in health food outlets and a few supermarkets.

Allinson
Bran Muesli
Bran Plus
Breakfast Muesli
Crunchy Bran with Cinnamon
Wheat Complete

Cheshire Wholefoods
Cheshire Wholefood Muesli

Granose
Crunchy Nut Cereal
Fruit Bran
Swiss Muesli

Prewetts
Wide range of additive-free cereals; below is a selection:
Fruit & Nut Muesli
Muesli Base

Shredded Bran
Sugar-free Muesli
Wholewheat Flakes

Real Foods
Breakfast Bran
F-Plan Diet Mix

Other brand names worth looking out for include Cotswold and Holly Mill.

Flour
Bleaches used in white flour do not have to be declared on labels.

Boots
Stoneground Wholemeal

British Home Stores
Self-raising Wholewheat   Raising agent sodium bicarbonate

Presto
Plain

Safeway
Strong Wholewheat

Sainsbury
Stoneground

Waitrose
Plain Wholewheat

Available through some supermarkets and health food outlets:

Allinson
Plain Wholemeal

Jordans
85% Plain
85% Self-raising   Raising agent sodium bicarbonate
Strong White – Unbleached
Strong Wholemeal

Prewetts
Rye
Wholemeal

Real Foods
These are generally only available in large quantities:
Unbleached White
Wholemeal Stoneground
Wholemeal Stoneground Special Bread
Wheatmeal

Other brand names to look out for are Doves Farm and Pimhill.

Bread

Boots
Country Grain Loaf	E300
Muesli Loaf	E300
Sesame Seed Roll	E300
Wholemeal Loaf	E300
Wholemeal Rolls	E300

British Home Stores
Bran Loaf	
Country Loaf	
Rye	
Multi-grain	E322

Marks & Spencer
Mixed Grain Loaf	
Brown Batched Loaf	E472(e)
Crusty Bread/Sesame Seeded	E472(e)
Crusty Brown Rolls	E472(e) △
Crusty Cob Rolls	E472(e)
Seeded Crusty Rolls	E472(e)

Tesco
White Pitta	E282 △

Waitrose
Wholemeal	E472(e) △

Goswell
Doves Farm Wholemeal – made with organically grown flour
Wholewheat

Prewetts
Wholemeal

All wholefood/health food outlets usually sell excellent additive-free bread.

Jams and Preserves

Boots
Range of honeys
Peanut Butter

Apricot Jam	$E_{440}(a)$
Blackcurrant Jam	$E_{440}(a)$
Cherry Jam	$E_{440}(a)$, E_{330}
Raspberry Jam	$E_{440}(a)$
Strawberry Jam	E_{440}, E_{330}
Thin Cut Orange Marmalade	$E_{440}(a)$

British Home Stores
Honeys

Conserves: Apricot	E_{330}
Blackcurrant	E_{330}
Raspberry	E_{330}
Strawberry	E_{330}
Blackcherry	E_{330}

Reduced Sugar Strawberry Jam
Reduced Sugar Orange Marmalade
Thin Cut Marmalade
Thick Cut Marmalade
Apple & Pear Spread

Fine Fare
Range of honeys
Range of 'Extra Jam' Conserves: Apricot
        Blackcurrant
        Ginger
        Raspberry
        Strawberry
        Swiss Black Cherry
        Three – Fruit Marmalade
Traditional Recipe Lemon Cheese (Lemon Curd-type preserve)
Peanut Butters

Marks & Spencer
Swiss Black Cherry Jam

Presto
Clear & Set Honey

Safeway
Honeys
Smooth & Crunchy Peanut Butter
A range of 'No Added Sugar Jams': Apricot
Blackcurrant
Marmalade
Raspberry
Strawberry

Sainsbury
Honeys
Smooth & Crunchy Peanut Butter

Tesco

Conserves:		
Apricot	E330
Blackcherry	E330
Blackcurrant	E330
Raspberry	E330
Strawberry	E330

Waitrose
Honeys
Peanut Butters

Mainly available in health food outlets:

Prewetts
Peanut Butter

Robertson
Sugar-free pure fruit spreads, various flavours

Rowses
Range of Honeys

Sunwheel
Peanut Butters
Sugar-free spreads, various flavours including Pear 'n' Apple
Sesame Spread
Sunflower Spread

Whole Earth
Peanut Butters
Sugar-free jams, various flavours

Other brand names to look out for include Ethos, Real Foods and
Thursday Cottage.

Fruit Juices
Almost all the main retailers now do own label fresh and longlife fruit
juices which are additive-free. Outlets include:
Bejam, Boots, British Home Stores, Co-op, Fine Fare, Marks &
Spencer, Presto, Safeway, Sainsbury, Tesco and Waitrose.
Other manufacturers producing additive-free fruit juices include:

Del Monte
Orange, Grapefruit, Apple, Pineapple

Lyons Maid
Juice bar range – pure fruit juices. (NB These have added sugar.)

St Ivel
Range of fresh fruit juices

Aspall
Apple juices and ciders

Other brand names to look out for in health food outlets include
Biota, Copella, Martlet, Prewetts, Volonté.

Margarine, Butter and Cooking Fats
Most of the large multiples' own label butters and suet appear to be
additive-free and these have not been listed individually.

Flora
Cookeen
White Cap
White Flora

Mainly available through health food outlets:

Granose
Red label margarine   E322

Other names to look out for include Vitaseig & Vitaquell.

Oils

All the major multiples have a range of cooking and vegetable oils which are additive-free, these have not been listed individually. Other manufacturers are listed below:

Flora
Sunflower oil

Brands mainly available through health food outlets include: Community, Prewetts, Real Foods, Sunwheel.

Cheeses

All the supermarket chains do a range of cheeses; many, especially the white cheeses, are additive-free. Some additives are commonly found in cheeses which include Double Gloucester E160(b), and Edam E160(b) and E251, Leicester E160(b). Cottage Cheese often has a number of additives, e.g. E412, E410, and E407. Below is a selection of additive-free cheeses from retailers.

Boots
Cheshire

Co-op
Cheddar, Cheshire, Derby, Lancashire

Marks & Spencer
Cheddar, Danish Blue, Leicester, Wensleydale

Presto
Natural & Matured Cheddar, Caerphilly, Derby, Lancashire.

Safeway
Brie, Blue Stilton, Cheshire, Goats Milk Cheese

Tesco
Brie, Cheddar, Cottage Cheese,
Lancashire

Waitrose
Brie, Camembert, Cheddar, Stilton

Loseley
Cottage Cheese

St Ivel
A range of white cheeses

Mainly available from health food outlets:

Prewetts
Cheddar, Cheshire, Lancashire, Wensleydale

Yoghurts

Fine Fare
Natural yoghurt

Marks & Spencer
Natural Yoghurt
'Low Fat' range of yoghurts free from
'artificial' flavouring and colouring   E163, modified starch

Presto
Hazelnut Low Fat Yoghurt   Modified starch
Natural Low Fat Yoghurt

Safeways
Natural
Natural Set

Sainsbury
Natural
A children's range of Mr Men Yoghurts
free from 'artificial' additives, but contain
several other additives. Seven flavours
including: Chocolate   E412 △, natural flavouring
          Banana   E100, E440(a),
                    natural stabiliser
          Fudge   E412 △, natural flavourings
          Strawberry   E162, E440(a)

Tesco
Natural Yoghurt

Loseley
A wide range of natural, fruit and Greek-style yoghurts

St Ivel
'Real' fruit yoghurts — six flavours, 'natural' flavouring

A wide variety of natural yoghurts from various farms are available from most health food outlets.

Soups

Boots
Packet soups: Mixed Vegetable with Spices
      Thick Green Bean
      Thick Potato

British Home Stores	
High Fibre Chicken	Malic acid, $\triangle$, natural chicken flavour
High Fibre Mushroom	Malic acid $\triangle$
High Fibre Tomato	
NB This is a 'no salt added' range.	

Fine Fare
Cream of Tomato

Tesco	
Tinned Tomato	Modified starch, E330

Heinz	
Cream of Tomato	Modified starch
Invaders	Modified starch

Available from health food outlets:

Prewetts	
Instant soup:	
Tomato, Lentil & Mushroom	Vegetable gums

Hugli
Range of packet soups

Fish Products
Most plain frozen and fresh fish are free from additives, but watch out for breaded or battered fish, fish cakes and fish fingers. Most tinned salmon, tuna, sardines and anchovies in oil and tomato sauce are also additive-free.

Listed below are those retailers which have other fish products which are free of suspect additives.

Bejam
By mid-1986, the following products should only contain E160(b) △ and E100. Check the labels.
Breaded Cod Steaks
Breaded Haddock Steaks
Fish Cakes
Fish Fingers

Marks & Spencer
Battercrisp Cod Portions
Crispy Breaded Haddock Fillets
Ovencrisp Cod Fillets

NB Read labels, e.g. Scottish Haddock
Fillet in Breadcrumbs contains E102, E124.

Sainsbury
Crispy Cod Portions
Crispy Haddock Portions
100% Cod Fillet Fish Fingers   E160(a), paprika

Tesco
Fish Fingers   Paprika, turmeric
but *not* those made from prime fillet

Meat Products

Bejam
Oven Ready Chickens free from polyphosphates
Premium Pork Sausages   E300

Marks & Spencer
Chicken Kiev
Chilli Con Carne (tinned)   Modified starch
Cornish Pasties   Modified starch
Crisp Bake Pork Pies
Duckling à l'Orange   Modified starch
Lattice Top Crispbake Pork Pie
Steak & Kidney pie, Traditional Natural Flavour

Sainsbury
Quarter Pounder Beefburgers

Meat Accompaniments

Fine Fare
Sage & Onion Stuffing

Presto
Parsley & Thyme Stuffing Mix
Sage & Onion Stuffing Mix

Sainsbury
Breadsauce Mix
Parsley & Thyme Stuffing
Sage & Onion Stuffing
Yorkshire Pudding & Pancake Mix

Mainly available at health food outlets:

Modern Health Products
Vecon – a vegetable stock paste

Canned Vegetables
Almost all frozen vegetables and chips are additive-free, exceptions being those that are breaded, battered or prepared in some way other than plain frozen. All retailers have a good range of frozen vegetables.
    The list below refers only to canned vegetables.

Co-op
Mixed Vegetables
Sliced Carrots
Whole Carrots
Italian Peeled Tomatoes

Fine Fare
Button Mushrooms
Sliced & Whole Carrots
New Potatoes
Mixed Vegetables

Marks & Spencer
Petits Pois
Sweetcorn
Unpeeled Jersey Potatoes
Baked Beans in Tomato Sauce   Modified starch

Ratatouille	Modified starch, E330

Also a range of prepared vegetable dishes:

Baked Potato with Cheese	
Broccoli in Cream Sauce	Modified starch
Cauliflower Cheese	Modified starch
Potato Croquettes	
Ratatouille	Modified starch
Spinach Mornay	Modified starch
Vegetable Bake	Modified starch, natural flavouring

Presto

Sliced Carrots	
Whole Carrots	
Mixed Vegetables	
Sliced Mushrooms	
Whole Mushrooms	E330
Sweetcorn	

Safeway

Baked Beans	Modified starch
Continental Mixed Vegetables	
Whole & Sliced Carrots	
Whole Green Beans	
Petits Pois	
Ratatouille	Modified starch

Sainsbury

Wide range including:

Broad Beans	
Cut Green Beans	
Whole & Sliced Carrots	
Sliced Mushrooms	
Pepper Salad	Modified starch
New Potatoes	
Ratatouille	Modified starch

Tesco

Beans in Tomato Sauce	Modified starch
Bean Salad in Wine Vinegar	
Carrots	
Whole Green Beans	
Mixed Vegetables	

Mushrooms
Petits Pois
Sweetcorn

Waitrose
Asparagus
Baked Beans in Tomato Sauce   E412 △
Carrots
Chick Peas
Mixed Vegetables
Mushrooms
Petits Pois
Ratatouille   Modified starch
Sweetcorn

Del Monte
A range of 'no salt added' vegetables

Heinz
Baked Beans in Tomato Sauce   Modified starch
Curried Beans with Sultanas   Modified starch,
                                     hydrolysed vegetable protein

Napolina
Baked Beans
Garden Peas
Green Beans
New Potatoes
Plum Peeled Tomatoes
Sweetcorn

Available through health food outlets:

Whole Earth
Baked Beans Campfire Style

Canned Fruit

Co-op
Wide range including: grapefruit, pears, peaches, prunes

Fine Fare
A range including: gooseberries, peaches, prunes

Marks & Spencer
Wide range including: grapefruit, pears, peaches, pineapple, satsu-mas

Presto
Mandarin Oranges in Light Syrup
Peach Halves & Slices in Syrup   E330
Pear Halves in Syrup   E330
Prunes in Syrup   E330

Safeway
Wide range including: peaches, pears, pineapple, mandarins

Sainsbury
Wide range of fruit in natural juices or fruit juice: apricot halves, blackcurrants, grapefruit segments, orange segments, peaches, pears, pineapple (*but not* raspberries in apple juice which contain E124). Also fruit in syrup: apricots, grapefruit, prunes, etc.

Tesco
Wide range including: grapefruit segments, peaches, pears in apple juice, pineapple, plums, prunes, satsumas

Del Monte
Range of fruit in natural juices

Napolina
Sliced Apples
Peach Slices & Halves
Pear Quarters & Halves

Biscuits, Cakes and Cake/Pastry Mixes

Boots
Wholemeal Apple Biscuits   Natural flavour
Wholemeal Fruit Bran Biscuits
Wholemeal Hazelnut Biscuits
Wholemeat Honey Biscuits
Wholemeal Muesli Biscuits
Hand baked biscuits: Coconut & Honey Natural flavour
                     Fig & Orange   Natural flavour
                     Lemon & Ginger Natural flavour
                     Sultana & Bran

British Home Stores
Wholewheat Bran   E322

Fine Fare
Digestives
Oatcakes Bran Enriched   Sodium bicarbonate
Oatcakes Traditional

Marks & Spencer
Digestives
Milk & Plain Chocolate Digestives
All Butter Thistle Shortbread
All Butter Shortbread Fingers
Cakes:
Country Cake
Fresh Cream Meringues
Chocolate Eclairs   E401, E471
Walnut Sandwich Cut Cake   Modified starch,
                                         natural flavouring

Presto
All Butter Shortbread Fingers, Assortment & Petticoat Tails
Savoury Biscuits:
Original Dutch Crispbakes
Wholewheat Dutch Crispbakes

Safeways
Meringues

Sainsbury
All Butter Shortbread Fingers & Rounds
Thistle Shortbreads
Wholemeal Shortbread
Peanut Crunch
Plain & Milk Chocolate Sweetmeal
Wholemeal Honey Sandwich

Tesco
All Butter Biscuits   Modified starch
Garibaldi
Honey Oatmeal
Scottish Shortbread – Assorted, Finger, Wholemeal Fingers

Spicy Fruit
Stem Ginger
Savoury Biscuits:
Cream Crackers
Poppy Sesame
Batter Mix

Waitrose
Shortbread Fingers & Shapes
Wholemeal Shortbread Fingers
Ginger Cookies — Sodium bicarbonate
Oatflake & Honey Cookies — Sodium bicarbonate
Tea Finger Biscuits — Sodium & ammonium bicarbonate

Treacle Cookies — Sodium bicarbonate

Crawfords
Balmoral Shortbread
Tartan Shortbread
Savoury Biscuits:
Butter Puffs (Cheese Biscuits)
Ryking Brown Rye
Ryking Light — E322, ammonium bicarbonate

Fox's
Natural Crunch Fruit & Nut
Natural Crunch Honey & Oat
Natural Crunch Choc Chip & Almond
Date, Nougat and Walnut Cookies
Sultana, Ginger & Lemon Cookies
Chocolate, Fruit & Nut Cookies
Coconut Cookies
Oatflake Cookies
Treacle Cookies
All Butter Shortbread
All Butter Sultana Cookies
Bran Crunch
Danish Shortcake
Muesli Biscuits
Wholemeal Honey Sandwich
Fivers

McVitie's
Abbey Crunch Creams
All Butter Shortbread
Digestive
Fruitcake
Oatcakes
Rich Tea
Royal Scot
'Natural Choice' range:
Blackcurrant Yoghurt Creams   Sodium & ammonium
                                     bicarbonate, E334

Fruit & Nut Crunch
Muesli Cookies
Wholemeal
Yoghurt Creams

Mainly available through health food outlets:

Allinson
Carob-coated Fruit & Nut
Carob-coated Ginger & Bran
Carob-coated Oatmeal
Hand-baked biscuits: Bran, Fruit & Nut, Ginger, Honey, Muesli,
Oatmeal, Walnut

                                     All contain sodium
                                     bicarbonate

Prewetts
Hand-baked biscuits: Fig, Sesame & Sunflower, Stem Ginger,
Sultana & Bran, Wholemeal

Ice-Creams

Sainsbury
Vanilla Ice-Cream:
Soft scoop, Choc Ices & Bricks   E410, E339 △, E466 △,
                                     E471, natural flavouring

Loseley
Dairy ice-creams, range of flavours E471, E401, E466 △, E33⬦

Desserts and Dessert Sauces

Marks & Spencer
Spotted Dick
Baked Apple Slice   E296 △, modified starch
Baked Rhubarb/Raspberry Slice   E296, modified starch
Rice Dessert & Apricot Purée   Stabiliser, natural
                                                              flavour
Syrup Sponge Pudding   E471, E415

Sainsbury
Chocolate Dessert Sauce
Chocolate Mint Dessert Sauce

Tesco
Apple Pie Filling

Only available through health food outlets:

Modern Health Products
A range of children's desserts due to be launched before Easter
which contain no artificial additives or preservatives. Six flavours:
blackcurrant & lemon jelly, carob dessert, pineapple, orange, blackcurrant, vanilla custard.

Convenience Meals

Bejam
A special range of additive-free meals include:
Beef Curry
Cauliflower Cheese
Chicken Curry
Chilli Con Carne
Lasagne
Spaghetti Bolognese

Boots
A vegetarian range:
Ratatouille
Risotto
Country Casserole   E340, natural flavour
Lasagne   E340

Fine Fare
Cheese & Tomato Pizza   E339, modified starch

Marks & Spencer
'Heat and Serve' range:
Cannelloni   Modified starch
Lasagne   Modified starch, E340
Ravioli   Modified starch, natural
                                          flavour
Creamy Mushroom Flan   Modified starch
Fresh Vegetable Flan   Modified starch
Quiche with Cheese & Onion   Natural flavour
Quiche with Tomato & Cheese   Modified starch

Sainsbury
Lasagne   Modified starch

Tesco
Cheese & Tomato Pizza   Modified starch
Cheese & Mushroom Pizza   Modified starch
Cheese & Tomato French Bread Pizza   Modified starch

Food Enterprises
A range of fresh chilled ready meals comprising of pasta and sauce
called 'Pasta Master' available in all the major multiples. Below are
some examples of the range (some contain edible starch but are
otherwise completely additive-free)
Cannelloni al Forno
Lasagne al Forno
Macaroni Cheese
Ravioli Napolitan
Spaghetti Bolognese

Available mainly through health food outlets:

Granose
A wide range of vegetarian meals in tins (some contain modified
starch and hydrolysed vegetable protein)

Höfels
A wide range of vegetarian meals in tins

Prewetts
A range of 'Heat and Serve' meals including:
Cannelloni
Tortellini
Ravioli
Wholemeal lasagne   Modified starch

Vegetarian Feasts
A wide range of 'Oven Ready' vegetarian meals which include:
Beans à la Grèque
Chilli Sin Carne
Macaroni Cheese
Vegetable Moussaka

Savoury Snack Foods

Fine Fare
Ready Salted Crisps

Safeway
Ready Salted Crisps

Sainsbury
Ready Salted Crisps
Twiglets

Tesco
Potato Shells

Waitrose
Ready Salted Crisps

Phileas Fogg
Corn Chips
Mignons Morceaux
Java Crackers
Tortilla Chips

Nabisco
Twiglets
Walkers Ready Salted Crisps

There are a wide variety of savoury snack products available in health food outlets. Below are listed two brand names, but there are many more.

Allinson
'Wheateats', various flavours   Carotene

Höfels
Soybrits

Chutneys, Sauces, Spreads

British Home Stores
Chutneys: Gooseberry   E260, modified starch
       Tomato & Apple   E260, modified starch
       Peach   E260, modified starch

Fine Fare
About to bring out a range of relishes free from 'artificial' colours,
flavours and preservatives in the following flavours:
Barbecue
Cucumber
Hamburger
Onion
Tomato & Chilli
Sweetcorn
Tartare Sauce   E415, acetic acid,
                                                natural flavour

Marks & Spencer
Tomato Ketchup   Modified starch

Safeway
Dressings – French, Italian & garlic oil
Mayonnaises – real, lemon & garlic   Natural flavour

Sainsbury
Brown Sauce   Modified starch, natural
                                                flavour

Chicken Paste   Natural flavour
Chicken Spread   Natural flavour
Cranberry Sauce
French Dressing
Italian Tomato Ketchup

Tesco
Apricot Chutney
Curried Fruit   Modified starch
Italian Garlic Dressing
Mustard Vinaigrette
Napoletana Sauce   E330, modified starch

Waitrose
Bolognaise Sauce   E330, modified starch
Chutneys: Apricot   Modified starch
         Mango   Modified starch
         Onion   Modified starch
         Tomato

Heinz
Chunky Piccalilli   Turmeric
Tomato Ketchup

Napolina
Tomato Ketchup
Napolitan Sauce
Tomato & Mushroom Sauce

A wide range of additive-free sauces etc. are available in health food
outlets. Below is a selection:

Down to Earth Wholefoods
Chutneys: Apricot & Date
         Lemon & Apple
         Mango & Orange
         Pineapple & Ginger
NB This range is also sugar and salt-free.

Granose
Sandwich spread – with Cereals
              with Herbs
              with Mushrooms
              with Olives

Tartex
Plain & Herb Patés

Vessen
Patés – Herb, Pepper, Mushroom

Whole Earth
Italian Sauce
Tomato Ketchup – Sugar-free
Traditional French Mustard

Confectionery

Boots
Coconut Crunch Bar
Oat & Honey Crunch Bar

British Home Stores
Carob Crunch Bars

Jordans

Original Crunchy Bars:	Apple & Bran	E322
	Coconut & Honey	E322
	Honey & Almond	E322

A very wide range of additive-free bars and confectionery is now available in all health food outlets.

Allinson
Carob Crunch Snack Bar
Carob Coated Sesame
Sesame Crunch

Granose

Fruit Snack Bars:	Apricot & Dates	E330
	Pear & Ginger	E330
	Strawberry	E330

Shepherd Boy
Banana Fruit & Nut Bar
Bran Fruit & Nut Bar
Coconut Fruit & Nut Bar
Ginger Fruit & Nut Bar
Sunflower Fruit & Nut Bar

Other names to look out for include Holly Mill, Prewetts and Sunwheel.

The Jargon Explained

FELICITY LAWRENCE with MELANIE MILLER and PETER MANSFIELD

Acids   Used to give foods a sharp or sour taste, especially soft drinks and confectionery. In jams, for example, they are added to counteract the overwhelming sweetness of the sugar, and to add to the flavour of the fruit. They are also used to preserve food, to dissolve colourings and to achieve the right level of acidity for other additives to function. Some, such as ascorbic acid (vitamin C, E300), are sold by manufacturers as 'added vitamins', but in fact these acids are also attractive for their other properties.

ADIs   Acceptable daily intakes. These are estimated safe levels of consumption for different additives, set by panels of experts (which often include representatives from industry) in the UK and by WHO/FAO. They are normally one hundredth of the level found to be toxic in animals. They can be something of a moving feast however. When the Ministry of Agriculture's Food Additives and Contaminants Committee found that the average child's consumption of yellow 2G (107) was more than five times the ADI, they decided that the safety limit must be wrong. For many additives, information is simply not available even to estimate safe levels or ADIs.

Allergy   To satisfy a doctor that they are suffering from an allergy, patients must not only show all the symptoms of hypersensitivity (see below), but also evidence of a measurable response in their immune system. This response includes changes in the composition and quantity of blood proteins, which occur when the body produces vast quantities of antibodies, and an increase in the numbers and reactivity of certain kinds of white blood cells involved in the immune process. Reactions must be reproducible every time exposure to the 'allergen', or substance causing the problem, occurs. And the reactions must happen whether the patient knows that he or she is being exposed to the substance or not.

   Few hypersensitive people meet all these criteria, and it may be that the majority are not allergic in this very narrow sense. But their

269

problems are superficially indistinguishable from allergy, and cannot be accounted for by 'suggestibility'. They deserve to be taken seriously and investigated with more commitment and enthusiasm by the scientific and medical fraternity.

Anti-caking Agents   These help prevent foods such as salt or icing sugar sticking together in lumps.

Anti-foaming Agents   These help stop food frothing, forming scum or boiling over during processing. In some cases they are also needed to control other additives: for example, glyceride oils are added to glucose syrup because it tends to foam during processing.

Antioxidants   When they come into contact with oxygen in the air, oils and fats are 'oxidised' and this makes them go rancid. Antioxidants prevent this happening. Unprocessed oils contain a natural antioxidant in the form of vitamin E, but this is destroyed in processing, and so manufacturers replace it with synthetic antioxidants. BHA and BHT (E320 and E321) are the commonest in this group, and both are highly suspect. They are often used not just to prevent oils from going rancid, but also to stop other additives, such as artificial colourings or flavourings, losing their strength – for example, in crisps. Because antioxidants give a longer shelf life, they also give bacteria more opportunity to grow, and so they may also be used in conjunction with preservatives.

Azo Dyes   Artificial dyes synthesised from diazonium compounds and phenol. The name is often used loosely to cover all artificial dyes. They are highly suspect – see numerical guide. Manufacturers are concerned at consumers' increasing awareness of the hazards of azo dyes and some are trying to promote a new name for them – 'the synthetic organic colours'.

Bases   Used to make foods less acid, or to help dissolve other acidic additives such as colourings. They may also react with acids to produce gases and so may be used as raising agents.

Bleaches and Improvers   Used to turn flour from its natural creamy colour to a 'whiter than white'. They are also used to speed up the maturing of refined flour for the Chorleywood Process – the high-speed method used to make three-quarters of the bread we consume in this country. They improve the raising ability of the flour. They also conveniently serve to sterilise the flour – stripping it of what little food value is left after refining, to the extent that not even weevils are interested in it, let alone humans.

Buffers   Chemicals which keep the acid/alkali balance of products constant during processing. Often necessary when manufacturers are adding large quantities of acids in the form of other additives.

Bulking Aids   Often used to eke out expensive ingredients or to give the impression that consumers are getting more food than they really are. Used most in slimming products, where they can make up between 10 and 20 per cent of the dry weight.

Carcinogen   A substance which can cause cancer.

Chelating Agents   Modern processing involves the use of machinery such as metal rollers and mixers, and as a result trace amounts of metal contamination are inevitably present in food. These contaminants are not only possible hazards, but may also make the food go rancid or change colour, so manufacturers add chemicals which 'chelate' or trap (literally claw, from the Greek) the metal atoms. The trouble is that chelating agents may also bind up vital minerals such as iron and prevent the body from using them.

Coal Tar Dyes   Artificial colourings synthesised from petroleum; most of them are highly suspect. (See also Azo Dyes.)

Colourings   Like dyes, substances used to replace colour lost during processing (because of heating or bleaching by other additives, for example), or to disguise the lack of real ingredients in food.

Concentrates   Ready mixed blends of additives for different sorts of food products, for example, concentrates made for 'fresh bread shops' contain emulsifiers, enzyme active soya, oxidants, salt, flavourings and maltol to give a fresh baked smell and flavour – guaranteed to pull customers in!

DHSS   The Department of Health and Social Security.

Diluents   Used to dissolve or dilute other additives.

Emulsifiers and Stabilisers   Emulsifiers are used to bind together water and fat, which usually repel each other; they are close relatives of soaps and detergents. Stabilisers stop the fat and water separating out once it has been mixed together. Manufacturers use them not only to bind fats in processed food, but also to keep bread soft, make cakes fluffy and to keep large quantities of fats added to processed meats from separating out. The polyphosphate group of emulsifiers is often used to make meat take up water, and so to increase its weight.

271

Enzymes　　Molecules which break down foods or help modify or synthesise them.

Excipients　　Additive powders used as carriers of other additives.

Firming Agents　　Used to make tinned and frozen vegetables and fruit retain its crispness, which would otherwise be lost in processing.

FAC　　The Food Advisory Committee, successor of the Food Additives and Contaminants Committee (FACC) and the Food Standards Committee (FSC), which advises MAFF on food policy. Its work includes reviewing additives in food.

FACC　　The Food Additives and Contaminants Committee, the predecessor of the FAC, which advised government on additives and contaminants in food until 1983.

Flavour Enhancers　　Also known as flavour modifiers, these are substances which have little or no flavour of their own but bring out the flavour of other substances, usually by stimulating the taste buds on the tongue. Often used these days to enhance not the natural flavour of food but the flavour of artificial flavourings!

Flavourings　　Synthetic flavours form the largest group of additives in use, and are vital to modern food processing. They are used to replace the natural flavour of food which may be lost in processing, to make up for a lack of real ingredients or poor quality food, and as substitutes for more expensive processes such as smoking. Synthetic flavours are preferred by manufacturers because they are more stable, so in response to consumer pressure for 'natural' additives industry has developed synthetic flavours which are 'nature identical', and so can be labelled as 'not artificial'.

Freezants　　Liquids or liquified gases, such as nitrogen, which extract heat from foods and freeze them by direct contact.

Free Radicals　　Free radicals are molecules which are reactive and unstable, and therefore tend to interact and interfere with any other molecules they meet in the body. Because they can disrupt cell metabolism, they are thought by some people to be the principal mechanism by which our bodies age and degenerate. Their role in disease is, however, still little understood.

Gelling Agents　　Used to thicken foods or make them gel.

Glazing Agents　　Waxes and oils such as liquid paraffin which are used to put a shine on food like sweets or dried fruit, or to add a protective gloss. Many EEC countries prohibit their use.

Humectants   Added to food to prevent it losing water and drying out.

Hyperactivity   Hyperactivity is a condition which can be triggered in children by eating certain additives, or other foods. It manifests itself as restlessness, unpredictable and unreasonable violence, lack of concentration, irritability and inability to sleep. Hyperactive children seem to have tireless stamina and often exhaust their parents. Their personality can be completely changed. Yet the whole condition can come in a matter of minutes, and go in a few hours. It has strong associations with other hypersensitive reactions. That the condition can be caused by food additives has never been fully accepted by the British medical establishment, though the view receives guarded support in the scientific literature. Many general practitioners, are, however, being forced by the evidence of their own eyes to take the condition seriously.

Hypersensitivity   Everyone is susceptible to irritation or damage from almost any substance, if they are exposed to enough of it; that is ordinary sensitivity. Some people acquire extreme sensitivity to one or several substances, to which they react severely even in quite tiny amounts. This hypersensitivity can be triggered by certain additives, other chemicals or foods, or it may be inherited. According to a report by the Royal College of Physicians and the food industry-funded British Nutrition Foundation, these reactions are rare and only affect a tiny number of people. But most medical practitioners would agree that hypersensitive reactions, clinically indistinguishable from allergy, are much commoner. Many practitioners also have the impression that more and more people are becoming hypersensitive to more and more substances. Reactions can include itching, running nose, eczema, etc.

Improvers   See Bleaches.

Intolerance   According to the medical establishment, most people who react adversely to additives do not suffer from what the average person would call allergy, but from intolerance. But the visible symptoms of intolerance are indistinguishable from those of allergy. See Allergy.

L-, D-, LD-, DL-   Some chemicals are not symmetrical and have a left or right-handed form. The left-handed form on its own is written as L-, as in L-glutamic acid; the right-handed form is written as D-. If both left and right handed forms are equally present, it is DL- or LD-. The notation gives you a useful clue to the chemical's origin. To get left or right-handed forms specifically, the substance would have to be produced biologically rather than synthesised.

MAFF   The Ministry of Agriculture, Fisheries and Food which is responsible, among other things, for food labelling and regulation of additives.

Mineral Hydrocarbons   Oils and waxes such as paraffins, derived from petroleum.

Mutagen   Mutations are changes in an organism's genetic material, which can be inherited. Agents which cause mutations are called mutagens. Mutations in cells can lead in some circumstances to the formation of cancer, and chemicals which are mutagens are often carcinogens.

Neurotoxic   Poisonous to the nervous system.

Packaging Gases   Gases used to fill packaging to prevent oxygen coming into contact with and damaging food.

Pesticides   A general term used to cover anything which kills any kind of pest. Insecticides kill insects, fungicides fungi, herbicides weeds, and so on.

PPM   Parts per million – the unit used to denote how much of a chemical is present in food. For example, dried apricots can legally contain up to 2,000 ppm (or 2,000 mg/kg) of sulphur dioxide. Some chemicals cause damage at levels as low as parts per billion in food. For example, the colouring ponceau 4R (E124) has been shown to provoke allergic reactions at this level.

Preservatives   Used to slow down decay caused by micro-organisms such as bacteria, yeasts and fungi, and to increase the storage life of foods. It is often argued that these additives are essential to protect consumers from dangerous substances such as salmonella and botulism, but there are several other ways of preserving food, such as refrigeration, freezing, better hygiene in factories, which are already widely available and used. The chemical characteristics which make preservatives effective in preserving food, also make them potentially harmful to health.

Propellants   Gases or highly volatile liquids capable of expelling foods from aerosol containers.

Releasing Agents   Used to prevent food sticking to packaging, moulds, tins, conveyor belts and other bits of machinery. Many releasing agents used in the UK are banned by other EEC countries.

Salicylates   Salts of salicylic acid, a substance chemically related to aspirin which occurs naturally in certain foods and can provoke allergic reactions. Those who suffer from asthma, hives or hyperactivity are particularly vulnerable. Those who are sensitive to salicylates frequently also suffer adverse reactions to some additives, particularly azo dyes and gallates.

Sequestrants   See Chelating agents.

Solvents   These are added directly to many foods as carriers for other additives, such as colourings, flavourings and emulsifiers, and they help them dissolve. They are also used to extract substances from raw materials. They will not necessarily be declared on labels.

Stabilisers   See Emulsifiers.

Synergists   Any additive which increases the effect of another is called a synergist, or co-worker. The antioxidants BHA and BHT (E320 and E321) are often used with citric acid, for example, because it enhances their effect.

Teratogen   From the Greek meaning monster, a teratogen is a substance which can cause birth defects, miscarriages and abortions.

Thickeners   Mainly gums, starches and celluloses used to thicken or stabilise the texture of foods. They may be used to add bulk to foods and make them look better value than they are.

WHO/FAO   The World Health Organisation and the Food and Agriculture Organisation.

Further Reading

The following are all recommended reading for the interested lay person:

L Doyal and S Epstein: *Cancer in Britain, The Politics of Prevention*, Pluto Press (London 1983)

J Elkington: *The Poisoned Womb*, Viking (Middlesex 1985)

M Hanssen and J Marsden: *E for Additives*, Thorsons (Wellingborough 1984)

P Mansfield and B Horner: *Look Again at the Label*, Soil Association (Suffolk 1984, available for 65p including post and packaging from 65 Colston Street, Bristol, BS1 5BB)

E Millstone: *Food Additives*, Penguin (London 1986)

M Miller: *Danger, Additives at Work*, London Food Commission (London 1985, available from London Food Commission, at £5, PO Box 29, London N5 1DU)

New Health magazine, Haymarket Publishing, Teddington, Middlesex (available from newsagents)

M Polunin: *The Right Way to Eat*, Dent (London 1978)

P Snell: *Pesticide Residues in Food: The Need For Real Control*, London Food Commission (London 1986, available as above)

C Walker and G Cannon: *The Food Scandal*, Century (London 1984)

The following are a small selection of the main specialist reference books and key American publications.

BNF: *Why Additives?* Forbes (London 1977)

J Burnett: *Plenty and Want: A Social History of Diet in England from 1815 to the Present Day*, Methuen (London 1983)

Commission of the European Communities: *Food Additives and the Consumer* (Brussels 1980)

D M Conning and A B G Lansdown: *Toxic Hazards in Food*, Croom Helm (London 1983)

E Cronin: *Contact Dermatitis*, Churchill Livingstone (Edinburgh 1980)

Food Act, 1984 (HMSO)

T E Furia: *Handbook of Food Additives*, CRC (Ohio 1980)

G G Gibson and P Walker (eds): *Food Toxicology, Real or Imaginary Problems*, Taylor and Francis (London 1985)

B T Hunter: *The Mirage of Safety*, Stephen Greene (Vermont 1982)

Industrial Aids Ltd: *Depth Study of Food Additives Industry* (London 1980)

International Agency for Research on Cancer: *Chemicals, Industrial Processes and Industries Associated with Cancer in Humans*, WHO, IARC supplement no. 4 (Geneva 1982)

International Agency for Research on Cancer: *Monographs on the Evaluations on the Carcinogenic Risk of Chemicals to Man* (Lyon, France)

M Jacobson: *The Complete Eater's Digest and Nutrition Scoreboard*, Anchor Press/Doubleday (New York 1985)

D J Jukes: *Food Legislation of the UK*, Butterworths (London 1984)

Leatherhead Food Research Association: *Food Additives*, 3rd ed. (1982)

Martindale (ed J E F Reynolds): *The Extra Pharmacopoeia*, 28th ed., Pharmaceutical Press (London 1982)

National Institute of Occupational Safety and Health: *Registry of Toxic Effects of Chemical Substances*, US Department of Health and Human Services (Cincinatti 1983)

M Pyke: *Food Science and Technology*, John Murray (London 1984)

R J Taylor: *Food Additives*, John Wiley (Chichester 1980)

J Verrett and J Carper: *Eating May be Hazardous to Your Health*, Anchor Books (New York 1975)

R Winter: *A Consumer's Dictionary of Food Additives*, Crown (New York 1984)

Reports of the Food Additives & Contaminants Committee
(FACC) – now Food Advisory Committee (FAC)

FACC Report Number
1. Antioxidants (Supplementary) 1965
2. Solvents 1966
3. Cyclamates 1966
4. Antioxidants (2nd Supplementary) 1966
5. Aldrin and Dieldrin 1967
6. Cyclamates (Supplementary) 1967
7. Further classes of additives 1968

8. Azodicarbonamide 1968
9. Emulsifiers and stabilisers 1970
10. Packaging 1970
11. Antioxidants (3rd Supplementary) 1971
13. Additives in Bread and Flour (FSC/REP/61, 1974 – Appendix 4) 1971
14. Preservatives 1972
15. Liquid freezants 1972
16. Emulsifiers and stabilisers (Supplementary) 1972
17. Solvents 1974
18. Antioxidants 1974
19. Liquid Freezants (Supplementary) 1974
20. Mineral Hydrocarbons 1975
21. Lead 1975
22. Flavourings 1976
23. Representation for the use of sulphur dioxide as an alternative to permitted colouring matter in canned garden peas 1977
24. Sorbic acid 1977
25. Solvents 1978
26. Beer 1978
27. Nitrites and Nitrates in Cured Meats and Cheese 1978
28. Flavour Modifiers 1978
29. Colouring Matter (Interim Report) 1979
30. Asbestos 1979
31. Modified Starches 1980
32. Bulking aids 1980
33. Infant formulae (FSC/REP/73, 1981 – Appendix 3) 1981
34. Sweeteners 1982
35. Enzymes 1982
36. Cheese (FSC/REP/75, 1982 – Appendix III) 1982
37. Cream (FSC/REP/76, 1982 – Appendix II) 1982
38. Metals in Canned Foods 1983
39. Arsenic 1984
FAC Report Number
1. Skimmed milk with non-milk fat regs 1984

Reports of the Food Standards Committee – now incorporated in the Food Advisory Committee
46. Canned Meat 1962
47. Meat Pies 1963
49. Fish and Meat Pastes 1965
52. Canned and Powdered Soups 1968

UK Regulations Controlling Additives

The Antioxidants in Food Regulations 1978 as amended by
The Antioxidants in Food (Amendment) Regulations 1980
The Sweeteners in Food Regulations 1983, and
The Bread and Flour Regulations 1984
The Chloroform in Food Regulations 1980
The Colouring Matter in Food Regulations 1973 as amended by
The Preservatives in Food Regulations 1974
The Colouring Matter in Food (Amendment) Regulations 1975
The Colouring Matter in Food (Amendment) Regulations 1976, and
The Colouring Matter in Food (Amendment) Regulations 1978
The Emulsifiers and Stabilisers in Food Regulations 1980 as amended by

The Emulsifiers and Stabilisers in Food (Amendment) Regulations 1982

The Sweeteners in Food Regulations 1983

The Emulsifiers and Stabilisers in Food (Amendment) Regulations 1983

The Cheese (Amendment) Regulations 1984, and

The Bread and Flour Regulations 1984

The Meat (Treatment) Regulations 1964

The Mineral Hydrocarbons in Food Regulations 1966

The Miscellaneous Additives in Food Regulations 1980 as amended by

The Food Labelling Regulations 1980

The Miscellaneous Additives in Food (Amendment) Regulations 1982

The Sweeteners in Food Regulations 1983, and

The Bread and Flour Regulations 1984

The Preservatives in Food Regulations 1979 as amended by

The Preservatives in Food (Amendment) Regulations 1980

The Jam and Similar Products Regulations 1981

The Preservatives in Food (Amendment) Regulations 1982

The Fruit Juices and Fruit Nectars (Amendment) Regulations 1982

The Sweeteners in Food Regulations 1983, and

The Bread and Flour Regulations 1984

The Solvents in Food Regulations 1967 as amended by

The Solvents in Food (Amendment) Regulations 1967

The Solvents in Food (Amendment) Regulations 1980

The Sweeteners in Food Regulations 1983, and

The Bread and Flour Regulations 1984

The Sweeteners in Food Regulations 1983

Contacts and Addresses

Food Additives Campaign Team, Room W, 25 Horsell Road, London N5 1XL. Please enclose a large sae.

The Hyperactive Children's Support Group will send further information about their work if you send 20p in stamps, and an sae (9″ × 4″) to The Secretary, 59 Meadowside, Littlehampton, West Sussex BN16 4BW.

The National Society for Research into Allergy has produced a directory of organisations (including local support groups) concerned about allergy. Send a large sae and 50p to them at PO Box 45, Hinckley, Leics LE10 1JY.

The London Food Commission provides research, education and advice on all aspects of food. If you would like to know more about their work write to the London Food Commission, PO Box 291, London N5 1DU.

Your MP can be contacted at the House of Commons, London SW1A 0AA.

The Minister for Agriculture, Fisheries and Food is the Rt Hon Michael Jopling MP, MAFF, Whitehall Place, London SW1A 2HH.

Food manufacturers' addresses are displayed on the packaging.

Index

INDEX